Distinguished Dissertations

Springer-Verlag London Ltd.

Other titles published in this Series:

Extensional Constructs in Intensional Type Theory
Martin Hoffman

Search and Planning Under Incomplete Information: A Study Using Bridge Card Play
Ian Frank

Theorem Proving with the Real Numbers
John Harrison

Games and Full Abstraction for a Functional Metalanguage with Recursive Types
Guy McCusker

Hardware Evolution: Automatic Design of Electronic Circuits in Reconfigurable Hardware by Artificial Evolution
Adrian Thompson

Models of Sharing Graphs: A Categorical Semantics of **let** *and* **letrec**
Masahito Hasegawa

Large Scale Collaborative Virtual Environments
Chris Greenhalgh

Architecture-Indepent Loop Parallelisation
Radu C. Calinescu

Randomized Algorithms: Approximation, Generation and Counting
Russ Bubley

User-Developer Coooperation in Software Development
Eamonn O'Neil

Inheritance Relationships for Disciplined Software Construction
Tracy A. Gardner

A Combination of Geometry Theorem Proving and Nonstandard Analysis, with Application to Newton's Principia
Jacques Fleuriot

Antonio Criminisi

Accurate Visual Metrology from Single and Multiple Uncalibrated Images

Springer

Antonio Criminisi
Dept. of Engineering Science, University of Oxford, 19 Parks Road,
Oxford OX1 3PJ, UK

Series Editor
Professor C.J. van Rijsbergen
Department of Computing Science, University of Glasgow, G12 8RZ, UK

British Library Cataloguing in Publication Data
Criminisi, Antonio
Accurate visual metrology from single and multiple
Uncalibrated images. - (Distinguished dissertations)
1.mensuration
530.8
ISBN 978-1-4471-1040-8

Library of Congress Cataloging-in-Publication Data
Criminisi, Antonio, 1972-
Accurate visual metrology from single and multiple uncalibrated images / Antonio Criminisi.
p. cm. -- (Distinguished dissertations, ISSN 1439-9768)
Includes index.
ISBN 978-1-4471-1040-8 ISBN 978-0-85729-327-5 (eBook)
DOI 10.1007/978-0-85729-327-5
1. Computer vision. 2. Geometry, Projective. I. Title. II. Distinguished dissertations
(Springer Verlag)
TA1634 .C75 2001
006.3'7--dc21 2001020903

ISSN 1439-9768

ISBN 978-1-4471-1040-8

http://www.springer.co.uk

Originally published by Springer-Verlag London Limited in 2001
Softcover reprint of the hardcover 1st edition 2001

Typesetting: Camera-ready by author

34/3830-543210 Printed on acid-free paper SPIN 10830685

"There are three branches of perspective:
the first deals with the reason of the diminution of objects as they recede from the eye, and is known as perspective of diminution;
the second contains the way in which colours vary as they recede from the eye;
the third and last explains how objects should appear less distinct in proportion as they are more remote.
And the names are as follows: linear perspective, the perspective of colour, the perspective of disappearance."

Leonardo da Vinci
Bibliothèque Nationale 2038 18r

Ai miei genitori, Elisabetta e Francesco

Preface

This monograph presents some research carried out by the author into *three-dimensional visual reconstruction* during his studies for the achievement of a Doctor of Philosophy Degree at the University of Oxford. This book constitutes the author's D.Phil. dissertation which, having been awarded the British Computer Society Distinguished Dissertation Award for the year 2000,[1] has kindly been published by Springer-Verlag London Ltd.

The work described in this book develops the theory of computing world measurements (e.g. distances, areas etc.) from photographs of scenes and reconstructing three-dimensional models of the scene.

The main tool used is *projective geometry* which forms the basis for accurate estimation algorithms. Novel methods are described for computing virtual reality-like environments from any kind of perspective image.

The techniques presented employ *uncalibrated images*; no knowledge of the internal parameters of the camera (such as focal length and aspect ratio) nor its pose (position and orientation with respect to the viewed scene) are required at any time. Extensive use is made of geometric characteristics of the scene. Thus there is no need for specialized calibration devices.

A hierarchy of novel, accurate and flexible techniques is developed to address a number of different situations ranging from the absence of scene metric information to cases where some world distances are known but there is not sufficient information for a complete camera calibration.

The geometry of *single* views is explored and monocular vision shown to be sufficient to obtain a partial or complete three-dimensional reconstruction of a scene. To achieve this, the properties of planar homographies and planar homologies are extensively exploited. The geometry of *multiple* views is also investigated, particularly the use of a parallax-based approach for structure and camera recovery. The duality between two-view and three-view configurations is described in detail.

In order to prove meaningful, measured distances must be associated with a measurement *accuracy*. Therefore, an uncertainty propagation analysis is developed to take into account the possible sources of error and the way they affect the uncertainty in the final measurements.

---

[1] `www.bcs.org.uk`

The general techniques developed in this book can be applied to several areas. Examples of feasible commercial, industrial and artistic applications are detailed in the book.

Redmond, WA, U.S.A.
April 2001

Antonio Criminisi

Acknowledgements

Many people have contributed in different ways to the success of this work and I am deeply indebted to all of them.

First of all I would like to thank my two supervisors Prof. Andrew Zisserman and Dr. Ian Reid. Their tireless fervour, encouragements and constant pursuit of the best possible solutions to the problems encountered have been fundamental in the achievement of the results presented in this book.

I am immensely grateful to my cousin and friend Prof. Roberto Cipolla. Without his passion for good science and his enthusiastic exhortations I would not have begun my studies in Computer Vision.

Many thanks to Prof. Martin Kemp. Working in a field where science and art can meet and cooperate has enriched my personal culture, improved my understanding of visual perception and turned on millions of "art-oriented" neurons in remote areas of my brain.

I am also indebted to Dr. Andrew Fitzgibbon for our inspiring conversations and his contribution to software development. David Liebowitz, with his passionate attitude towards geometric problems, has been of great help in creating many three-dimensional models and animations as well as proofreading part of this work. Philip Pritchett, Frederik Schaffalitzky, Geoffrey Cross, David Capel, Richard Smith, Caroline Bailard and Karen McGaul in the Visual Geometry Group have all contributed to enriching my understanding of Projective Geometry. Thanks to Lourdes, Josephine, Aileen and Claudia (Portuguese) for bearing to see my face at home as well as in the office every day; they never complained. Thank you Josephine for proofreading.

I would also like to thank A. Blake, L. van Gool, M. Brooks, D. Murray, P. Anandan, S. Carlsson, R. Hartley, P. Torr, J. Mundy, J. Krakora, J. Levenson, C. Wong and P. Galluzzi for contributing in different ways to the development of many parts of this book.

In the development of this work I have also been supported by many friends who, especially at the beginning of my stay in England, have helped me to overcome any language problems and enjoy my time with them. Among them I would like to thank Claudia (Mexicana), Zam, Frank, Rory, Arnd, Francesca, Ronan, Anna, Hester, Arnoud, Mark (another victim of the proofreading disease), Rebekah, Yaeli and little Michael, Liz, Graham, Tobias and many, many others. Believe it or not, going to the pub and on holiday with them has had extremely positive effects on the quality of my work. Many thanks also to the people who have agreed to appear in pictures in this book.

I should not forget my many Italian friends, colleagues and professors scattered all over the world, always ready to provide suggestions and help.

Huge thanks to my parents Elisabetta and Francesco, my sister Carmela, Nicola, Stefano, Giovanna, Giacomo and Costanza. From the opposite side of the continent they have always supported my choices and my ideas and provided me with lots of love.

Dulcis in fundo, zillions of thanks to my wife Beatrice (and the little baby growing inside her), with her love and caring support she has made possible the realization of this book.

Grazie!

Contents

1. Introduction

Images or sequences of images potentially carry a tremendous amount of geometrical information about the scene represented. The aim of the work presented in this book is to extract this information in a quantifiable, accurate way.

1.1 Accurate measurements from images

General techniques are developed which find their engineering motivation in the realization of a flexible, fast and robust visual measuring device, capable of reconstructing a three-dimensional digital model of a scene from interpreting photographs (see Fig. 1.1). The principal idea is:

- an operator takes one or more photographs of the scene (or object) to be measured;
- a computer, by interpreting those images, creates a three-dimensional metric model of the viewed scene [5];
- the model is stored in a database which may be queried at any time for measurements via a graphical user interface.

Such a device possesses several interesting features. (i) It is user friendly. In fact, once the images are taken and the model built, an operator can virtually walk through it, view the scene from different locations, take measurements querying the software interface and store them in a database (***visual metrology***), interact with the objects of the scene, place new, consistent virtual objects in the scene (***augmented reality***) and create convincing animations. (ii) The capture process is rapid, simple and minimally invasive since it only involves a camera to take pictures of the environment to be measured. (iii) The acquired data are stored digitally on a disk ready for reuse at any time negating the need to go back to the original scene when new measurements are needed. (iv) The hardware involved is cheap and easy to use. No new, dedicated hardware is necessary.

All the work presented here is motivated by the necessity of understanding the mathematical theory underlying such a visual measurement device.

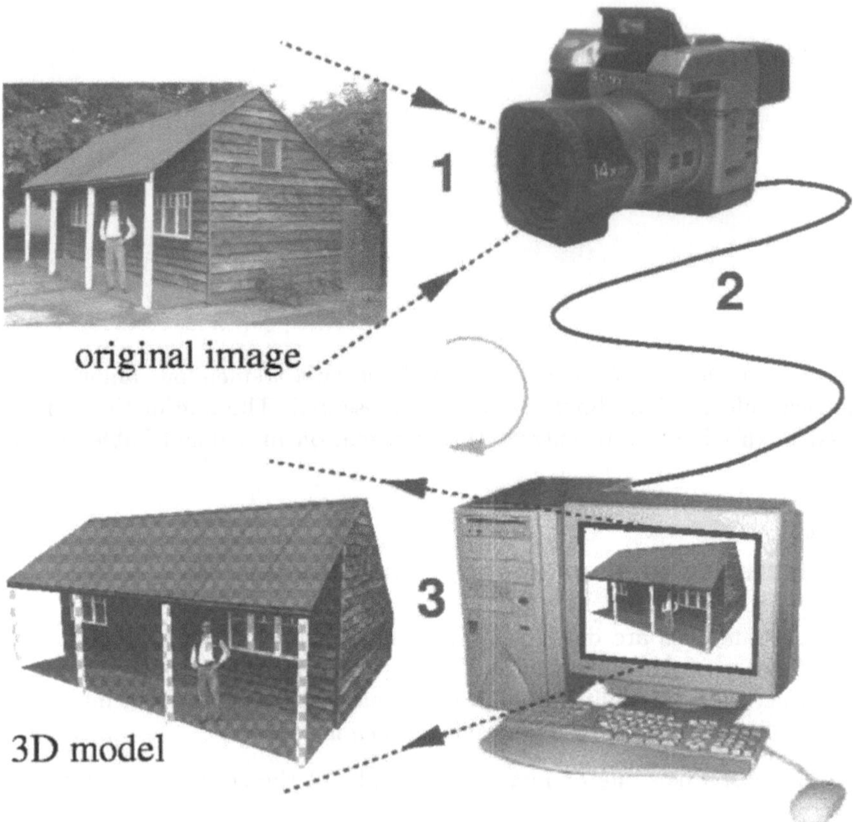

Fig. 1.1. A three-dimensional visual measuring device: a photograph of a scene is taken; the image is transferred into a computer; the computer reconstructs the geometry of the viewed scene in a three-dimensional digital model. Once constructed the model can be interactively queried for measurements.

The mathematical basis.
The theory developed in this book proceeds from a strong and reliable mathematical basis, *projective geometry* [38, 58, 106]. In particular, 2D-2D homographic transformations and more general 3D-2D projectivities are investigated and used extensively. While much of the underlying theory already existed, in order to achieve the goal of this book a number of research areas had to be addressed.

The algorithms developed require no knowledge of the camera's internal parameters (focal length, aspect ratio and principal point) or external ones (position and orientation); i.e. no internal or external camera calibration is needed. Camera calibration is replaced by the use of scene constraints (often referred to as *scene calibration*) such as planarity of points and parallelism of lines and planes. These geometric cues are inferred directly from the images; no scene markers or specialized sensors are required.

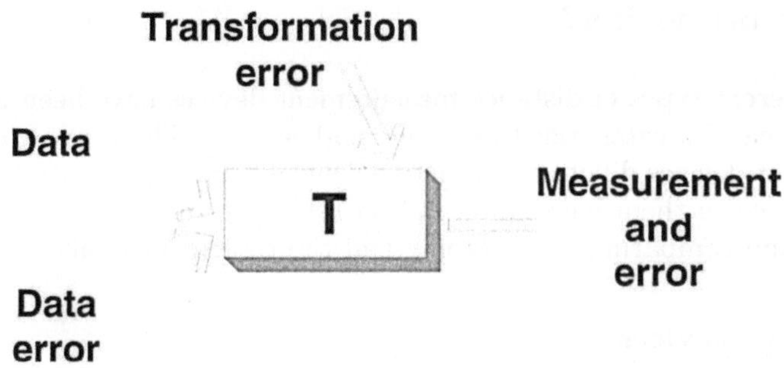

Fig. 1.2. Error propagation: the input data are processed by the transformation **T** to obtain the required output measurement. Input data and transformation are affected by error, leading to error in the output measurement

A hierarchy of novel geometric techniques to calculate distance measurements is investigated to address a range of different cases. The techniques span situations where no metric information is known about the world (*completely uncalibrated camera*), through to cases where some reference distances are known but they are not sufficient for a complete camera calibration (*partial calibration*). This leads to extremely flexible algorithms which can be applied to a wide range of images such as: photographs of buildings and interiors, aerial images, archived images, photographs of crime scenes and even paintings.

Both single- and multiple-view configurations are employed. *Stereo vision* has in the past been used to compute depth but, in this book, *monocular vision* proves to be sufficient, in many cases, to obtain partial or complete three-dimensional reconstruction. The use of multiple views in a *parallax*-based framework is also considered in cases where single-view metrology cannot be applied.

Accuracy of measurements.
The process of taking measurements is traditionally an engineering task and, like all engineering tasks, must be accurate and robust. Scene measurements are obtained from input data (e.g. image points) by a geometric transformation (e.g. a homography). Data and transformation are affected by error and so is the output measurement (see Fig. 1.2). A proper treatment of error and its propagation through the chain of computations is therefore necessary.

The uncertainty analysis developed in this book takes into account all the possible sources of error (the accuracy of the device used, the error affecting the data acquisition process and the operator) and predicts how these affect the accuracy of the final measurement. Thus an uncertainty estimate is associated with each output measurement.

This is achieved by using a *first order error propagation*. But first order analysis involves a linear approximation of the non-linear transformations relating input data to output measurements. Therefore the validity of the approach has to be assessed. This is done by employing *Monte Carlo* statistical tests.

1.2 Why use vision?

Several different types of distance measurement devices have been used in the past. They can be categorized as *active* and *passive*. These two categories of measurement devices differ in that active devices send signals into the environment and receive them back. Information related to distances is retrieved by analyzing and comparing the outgoing and the returned signals.

1.2.1 Active devices

Ultrasonic devices.
Many distance measuring systems have been based on ultrasonic technology. It is possible to buy relatively cheap ultrasonic devices capable of measuring the distance of the operator from an object (such as a wall), relying on an echo reflection time measurement system.

Ultrasonic scanners have, for instance, been successfully used in medical imaging for three-dimensional reconstruction of the structure of bones [119] or other internal tissues. They have also been employed in robotics problems such as autonomous vehicle navigation [48], where the accuracy and speed of the localization system are vital. The main problem with such an approach is that the measurement returned is affected by poorly understood and almost unpredictable phenomena like multiple reflections of the ultrasound waves on various objects, thus leading to wrong estimates of the reflection time.

Laser range finders.
A second approach for measuring depths is the use of laser range finders. These devices work by directing laser beams onto the object to be measured and analysing the phase or echoing time of the reflected beams.

Laser range finders are being successfully applied to metric shape reconstruction of relatively small objects [8]. For instance, they have been employed in the Michelangelo Project [76] to create three-dimensional models of ancient statues such as Michelangelo's *David* in Florence. Laser devices have also been applied in autonomous robot navigation [63]. Those systems are extremely accurate but they suffer problems similar to those of ultrasonic devices. Laser-based devices are usually tuned to detect a specific range of distances and they can be quite expensive and, sometimes, dangerous.

Structured light.
Other active devices employ cameras to acquire images of an object illuminated by a regular light pattern. Some auxiliary devices (a laser emitter or just a slide projector) project a light pattern or a set of patterns onto an object. The shape of the object is computed from the deformation of the projected grid.

Structured light-based approaches have been used for accurate measurement of surfaces of close-range objects which do not possess enough texture [81]. Those methods have also been used to capture facial expressions [35, 97]. The need for auxiliary light projective devices leads to a loss in the flexibility of the measuring tool.

In general, the measurements obtained from all active devices can be badly affected by unexpected reflections or interferences and therefore need to be used with extreme care. Furthermore, such devices can only be applied to measure objects which are easily accessible, i.e. applied in situ. For instance, they are no use to measure scenes of which only images exist.

1.2.2 Passive devices

Passive devices such as cameras do not suffer from the above problems and are characterized by a wider range of application.

They can be applied to measure the distance of the device from an object (like range finders) as well as the distance between two other points in space, areas of surfaces and angles. Speed is not an issue for such devices.

Cameras return bi-dimensional data (rather than mono-dimensional ones) characterized by dense sampling within the field of view. They do not need to be tuned to a specific range of action but can measure distant objects as well as close ones. The accuracy of the measurements will depend though on the relative distance between the camera and the object to be measured, as will be investigated in detail throughout this book.

Furthermore, since images are the only input data, visual metrology can successfully be applied to reconstructing objects from archived images, old footage, postcards and paintings (unlike active devices).

1.3 Why is visual metrology hard?

The previous section has described the advantages of using cameras as opposed to other measuring devices. However, taking measurements of the world from images is complicated by the fact that in the imaging process the three-dimensional space is projected onto a planar image, with some unavoidable loss of information. Reconstructing the scene means retrieving that information from the images.

In particular, perspective distortions occur during the acquisition stage. For instance, objects which are far from the eye (or camera) look smaller than objects which are close[1]. Examples of perspective distortion on real images are shown in Fig. 1.4.

In the past, photogrammetrists have addressed the problem of establishing real measurements from images. Their techniques rely on knowing the internal parameters of the camera, its exact position with respect to the viewed scene or the position of marker points in the scene (*georeferenced images*). Unfortunately, photogrammetric algorithms work well only if those values are known with great accuracy, and even a very small deviation can cause large errors

[1] This basic intuition was formalized first by Euclid in his *Optica* [36] and then employed by mathematicians and artists in the Italian Renaissance, among which Leonardo da Vinci is one of the outstanding examples (see Fig. 1.3).

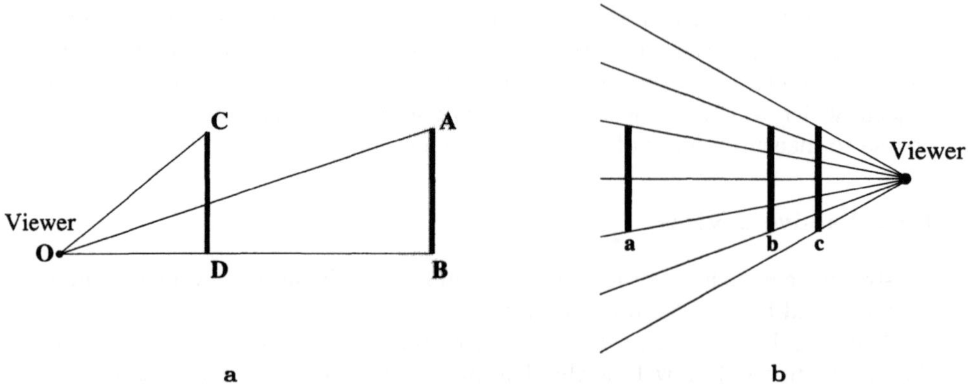

Fig. 1.3. First proofs of perspective effect: **a** Euclid's proof (ca. 300 BC); **b** Leonardo's proof: *"Among objects of equal size that which is most remote from the eye will look smallest."* Leonardo da Vinci (1452-1519), Codex Foster II 15v.

Fig. 1.4. Perspective distortion in the image acquisition process: **a** A photo of a wall of Keble College in Oxford: the windows are foreshortened differently depending on their distance from the camera. **b** The four pillars have the same height in the world, although their images clearly are not of the same length because of perspective effects.

in the final measurements. Furthermore, camera internal parameters are quite sensitive to changes of temperature and mechanical shock. Photogrammetric techniques cannot be applied to archived images and have no application for paintings.

In this book, projective geometry is employed in a flexible way to model the perspective distortion and recover full or partial three-dimensional information from uncalibrated images and geometric cues inferred directly from the images.

1.4 Applications and examples

The theory presented here can be applied to solving many of the engineering and architectural problems which involve measuring objects. For instance, it

a b

Fig. 1.5. Measuring distances and estimating their accuracy: **a** original image, a photograph of the facade of the Crystallography Laboratories in Oxford; **b** the height of different windows has been computed and the measurement accuracy estimated.

is useful to photogrammetry researchers for measuring dimensions of buildings from aerial images as well as in video compression techniques. In fact, if a three-dimensional model of a scene is known, then it is possible to eliminate all the unnecessary and redundant information and obtain a high-rate compression of the data space to be stored or transfered.

This section analyses three of the many possible uses, which will be described in detail in the following chapters.

1.4.1 Architectural and indoor measurements

An important application is in taking measurements of man-made scenes (measuring objects such as: furniture, doors and windows).

Geometry-based approaches are used here to compute distances on planar surfaces [24] (details in Chapter 4). For example, the edge of a window can be transformed from the image into the three-dimensional world by a geometric transformation, thus computing its real length (see Fig. 1.5). The accuracy of the measurements is also estimated.

Perspective distortions on planar surfaces are therefore modelled correctly and rectified (*fronto-parallel*) views created (see Fig. 1.6). Once two images of a world plane have been corrected for projective distortion, they can also be stitched together to make a mosaic image (see Fig. 1.7).

Furthermore, using the metric information retrieved for each planar surface and combining it with that computed for the others, an indoor environment can be modelled as a simple shoe-box room. Protrusions from planar surfaces (columns, pipes and windows sills) can be measured using two or more views [23] (details in Chapter 6).

1.4.2 Forensic measurements

An important application of this theoretical framework is in forensic science, to measure dimensions of objects and people in images taken by surveillance cameras [28].

a b

Fig. 1.6. Rectification of a planar surface: **a** original image, Keble College, Oxford; **b** rectified image. Perspective distortion has been removed; now the image appears to be front-on. The windows which have the same size in the world now have the same size in the image too.

a b c

Fig. 1.7. Mosaicing two images: **a** and **b** two images of The Queen's College Middle Common Room in Oxford; **c** mosaic image. The images have been perspectively corrected and stitched together.

Because of the very bad quality of the images (taken by cheap security cameras), quite often it is not possible to recognize the face of the suspect or distinct features on his/her clothes. The height of the person may become, therefore, an extremely useful identification feature. The main problem is that

a b

Fig. 1.8. The height of a person standing by a telephone box is computed: **a** original image; **b** the computed height of the person and the estimated uncertainty are shown.

the suspect has usually left the scene of crime or the scene has changed (it might have been blown up); therefore, in order to compute his/her height we can rely only on the actual image. This is complicated by the fact that usually no more than one view of the scene is available.

Chapter 5 will show how to solve this problem. It describes techniques to estimate the heights of people from a single uncalibrated view (see Fig. 1.8). Good results have been achieved in this field, some of which have already been applied successfully by the forensic science community [28].

1.4.3 Reconstructing from paintings

Much interest has also been expressed in the possibility of building colourful three-dimensional models and creating beautiful animations of painted scenes, so that the observer can "dive" into the painting itself [24, 77].

The mathematical theory developed in this book is particularly applicable to paintings and drawings which follow the geometric rules of *linear perspective*.

Linear perspective was invented at the beginning of the fifteenth century in Florence by the Italian architect Filippo Brunelleschi [67] and mathematically formalized by Leon Battista Alberti in 1435 [1]. Immediately after its invention, linear perspective was applied by the masters of the Italian Renaissance (e.g. Masaccio, Piero della Francesca, Leonardo and Raffaello) to paintings, drawings and engravings, providing them with new expressions of depth and space.

A painting which follows the geometric rules of perspective behaves as a *perspective image* (like a photograph) and therefore the projective techniques described in this book can be successfully applied in such case (see Fig. 1.9). However, a preliminary check on the correctness of the perspective (location of vanishing points, perspective effect etc.) becomes necessary.

a b

Fig. 1.9. Three-dimensional reconstruction from a painting: **a** *St Jerome in His Study* (1630), by Hendrick V. Steenwick (1580–1649), private collection (Joseph R. Ritman Collection), Amsterdam, The Netherlands; **b** a view of the reconstructed three-dimensional model. Notice the reconstructed, beautiful window.

1.5 Summary

1.5.1 Contributions

This book improves on the state of the art on various aspects of computer vision, understanding of photographs and visual art:

- A hierarchy of novel, accurate and flexible techniques is developed to make measurements and reconstruct three-dimensional scenes from uncalibrated images. They range from cases where no information about the scene is known to situations where only partial calibration is available.
- Affine and metric reconstruction is made possible from single views as well as pairs or sequences of images.
- An analysis of the accuracy of the reconstruction is developed which predicts how errors affect the final measurements. The analytical theory is validated by means of statistical tests.
- The general algebraic techniques developed here open new ways of interpreting and understanding classical and modern works of art.

1.5.2 Book outline

Chapter 2 starts with a literature survey of the most relevant research conducted in the fields of video metrology, three-dimensional reconstruction, error analysis and art history.

In Chapter 3 some background material is presented: an introduction to the basic geometry employed (3D-2D and 2D-2D projective mappings); the notation used; fundamental image processing algorithms (edge detection and radial distortion correction); and the basics of error propagation theory.

Chapter 4 brings the underlying mathematics to life by developing algorithms to perform metrology on planar surfaces. Different methods of accurately estimating ***plane-to-plane homographies*** are investigated. Planar homographies are then used to measure distances between points or parallel lines on a world plane. In order to estimate uncertainties in those measurements a first order error analysis has been developed, which takes account of all major sources of error. A simple working application is also presented.

Chapter 5 extends measurements on planar surfaces to the three-dimensional case. Algorithms for recovery of partial and complete three-dimensional structure from single uncalibrated images are explored. The techniques make extensive use of planar projective transformations like ***homographies*** and ***homologies***, thus avoiding traditional problems like matching corresponding features over multiple views of a scene. Particular attention is paid in this chapter to: (i) measuring distances between two parallel planes; (ii) measuring on parallel planes and comparing these measurements to those obtained in any other parallel plane; and (iii) determining the camera position. Uncertainties on the computed measurements are also estimated. This work is generalized to a parametrization of the three-dimensional space as a collection of three pencils of parallel planes in three different directions.

Chapter 6 deals with multi-view geometry and describes the advantages of this approach over the single view. The geometry of two and three views gives rise to an elegant algebraic description of the three-dimensional space, employing planar homologies. It is shown that homologies encapsulate the fundamental geometric constraints: the epipolar constraint in the two-view case and the structure constraint in the three-view configuration. The elegance of this approach is highlighted by the duality relationship which arises naturally between the two-view and three-view configurations. Furthermore, a new interpretation of the space based on a ***plane-plus-parallax*** approach leads to simple formulations for computing structure and camera location.

Several examples on real images are provided in each chapter. Further examples are presented in the colour plates in Chapter 7. They show images of the models computed by applying the described techniques to photographs and paintings.

Chapter 8 concludes the book with: a summary of the presented problems and the methods to solve them; a discussion of the advantages and disadvantages of employing such approaches; a list of possible ways to improve the analyzed techniques and future directions of research.

2. Related work

This chapter presents a survey of the most significant work in the field of three-dimensional reconstruction from two-dimensional images, and uncertainty estimation.

Using images for measuring and reconstruction.
Visual metrology and three-dimensional reconstruction of scenes from images have been, in the past few years, amongst the main aims of computer vision. Much effort has been put into pursuing such goals, and three main difficulties have been encountered:

- reconstructing from multiple images fundamentally involves solving the "correspondence" problem, i.e. which set of features in the images is genuinely the projection of a feature in the scene;
- a reliable and well-defined geometric structure of the scene needs to be seen if a single-view approach is applied;
- errors inevitably propagating along the computation chain cause a loss of accuracy of the measurements and thus the final structure.

Several researchers have been interested in *Euclidean* or more simply *affine* or *projective* reconstruction. Most of them have employed multi-view approaches but only a few have investigated the problem of conducting a proper uncertainty analysis to assess the accuracy of the final structure.

In this chapter different ways of recovering structure by using one or more views are analyzed, as well as various methods to estimate the associated uncertainty.

The articles considered range from single-view systems to bi- and multi-view ones. Work related to parallax-based approaches, the estimation of accuracy in metric reconstruction and geometric analysis of perspective paintings is also examined.

2.1 Using one view

In general, one view alone does not provide enough information for a complete three-dimensional reconstruction. However some metric quantities can be computed from the knowledge of such geometrical information as the relative position of points, lines and planes in the scene. But in order to achieve this,

Fig. 2.1. Tsai grid: camera parameters can be calibrated from the known pattern on the grid.

in general the intrinsic parameters of the camera need to be known. These are *focal length*, *principal point*, *skew* and *aspect ratio* [38].

A number of visual algorithms have been developed to compute the intrinsic parameters of a camera from images in the cases where they are not known. This task is called *camera calibration*. Usually, calibration algorithms assume that some of the internal parameters are known, and derive the remaining ones. Common assumptions are: unit aspect ratio, zero skew or coincidence of principal point and image centre. The calibration techniques described in this section are based on a single-view approach.

The work of Tsai [126] has been particularly influential in the field of camera calibration. From a single image of a known, planar calibration grid (see Fig. 2.1) it estimates the focal length of the camera and its external position and orientation assuming known principal point. An attempt to analyze the accuracy of the estimated parameters has also been reported.

Caprile and Torre, in their classical work [12], develop an algorithm to compute the internal and external parameters of the camera (they assume unit aspect ratio and zero skew) from single views. They make use of simple properties of vanishing points; these can be extracted directly from the image by intersecting images of parallel lines. A simple calibration device consisting of a cube with sets of parallel lines drawn on its faces is employed. In particular the authors demonstrate that the principal point of the camera coincides with the orthocentre of the triangle whose vertices are the three vanishing points for three orthogonal directions. An earlier work on this subject can be found in the photogrammetry literature [51].

The problem of calibrating a camera is discussed in [38] by Faugeras. He presents algorithms to compute the projection matrix (*external calibration*) and eventually the internal parameters of the camera from only one view of a known three-dimensional grid. He analyses linear and non-linear methods for estimating the 3D-2D projection matrix, the robustness of the estimate and the best location of the reference points.

In Liebowitz and Zisserman's work [78] camera *self-calibration* is obtained simply from images of planar structures like building facades or walls, with distinguishable geometric structure. Use is made of scene constraints such as

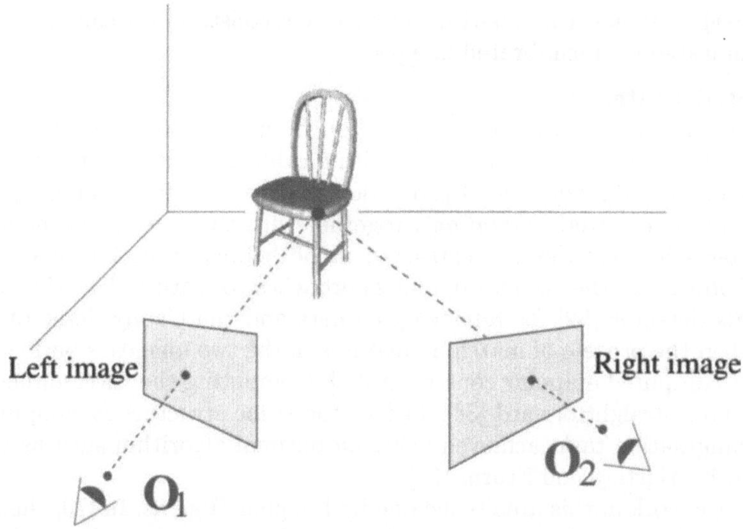

Fig. 2.2. Stereo acquisition system scheme: two images of the same scene are captured from different viewpoints ($\mathbf{O}_1$ and $\mathbf{O}_2$). Three-dimensional structure can be computed by comparing those images.

parallelism and orthogonality of lines and ratios of lengths. No specifically designed calibration object is required.

An interesting problem is addressed in [68] by Kim *et al.* In this paper the authors compute the position of a ball from single images of a football match. By making use of shadows on the ground plane and simple geometric relationships based on similar triangles, the ball can be tracked throughout the sequence.

This book develops a new approach to the reconstructruction of three-dimensional scenes from single images which does not need a full camera calibration (Chapter 5).

2.2 Using two views

The classical algorithms for three-dimensional reconstruction use *stereo vision* systems [37]. Stereo vision consists of capturing two images of a scene, taken from different viewpoints, and estimating the depth of the scene by analyzing the disparity between corresponding features (see Fig 2.2). This methodology finds its basis in trigonometry and triangulation and is employed by the human binocular vision system.

The basic steps in reconstructing a scene from two images are:

1. finding corresponding points on the two images.
2. intersecting the corresponding rays in three-dimensional space.

The following shows a comparison of how a reconstruction can be achieved using calibrated or uncalibrated images.

Calibrated route.
The typical route for structure computation from pairs of calibrated images is as follows. Once two images of a scene with a calibration grid in it are captured by a fixed stereo rig, the 3D-2D projection matrices for each camera [44, 125] are computed (*external calibration*), together with the *epipolar geometry* [38]. Some interesting features are extracted in both images, mainly corners and edges of objects in the scene. Popular approaches to feature detection include the Harris detector [52] for retrieving corners and the Canny detector [9] for edges. Then the process of matching features in the two images is performed by using the computed epipolar constraint [93]. Computing the three-dimensional depth is now straightforward [38]. In fact the scene structure is computed via a ray triangulation task, achieved using an optimal algorithm such as the one described by Hartley and Sturm [57].

A classic work in this area is the one by Longuet-Higgins. In [79] the author presents an algorithm for computing the three-dimensional structure of a scene from a pair of perspective projections. If the images of eight world points can be located in both projections, then the three-dimensional location of other points and the relative geometry between the two viewpoints are obtained by solving a set of eight simultaneous linear equations. A fundamental application of this result is in the reconstruction of a scene from two calibrated views.

A method for three-dimensional reconstruction based on homography mapping from calibrated stereo systems is described in [138]. Zhang and Hanson investigate the problem of recovering the metric structure of the scene by mapping one image into the other using homographies inducted by planar surfaces in the scene. Using at least four coplanar correspondences, the three-dimensional structure can be achieved in Euclidean space up to a scale factor (*scaled Euclidean structure*) and two real solutions. In order to disambiguate the two solutions a third view is required.

However, using a calibrated stereo rig has some disadvantages since the camera internal parameters may not always be constant. It might be necessary to adjust the focal length, or accidental mechanical and thermal events can affect it. In order to overcome those problems, new algorithms for computing three-dimensional structure from images with unknown camera calibration have been developed in the past few years (uncalibrated route).

Uncalibrated route.
If the two cameras used during the acquisition process are uncalibrated (unknown internal parameters), then only a *projective structure* can be obtained [37]. Corresponding points on the two images can be matched and thus the *epipolar geometry* estimated. The two projection matrices are computed from the *fundamental matrix* but only up to a projectivity.

However, in order to use a pair of cameras as a world measuring device a complete metric reconstruction is required. In order to extend a projective structure to an Euclidean one, some additional geometric information about the viewed scene is necessary. Several algorithms have been developed [39] to deal with this problem.

Given a pair of uncalibrated images, the two cameras can be calibrated from the analysis of the point matches in the two views. A self-calibration, non-iterative algorithm, based on matrix factorization has been developed by Hartley [53]. In this work the camera intrinsic parameters are assumed to be constant. They are all known apart from the focal length which is computed, together with the relative pose of the cameras. No specialized calibration device is necessary.

Koenderink and van Doorn in [69] introduce the notion of *affine structure* from two uncalibrated views. The authors present a method for recovering structure from motion in a stratified manner. The results can be reformulated in terms of motion parallax.

In [87] Moons *et al.* present an algorithm for recovery of three-dimensional affine structure from two perspective views taken by a camera undergoing pure translation (equivalently: translating object and stationary camera). The images of five points in two views need to be seen. The algorithm is further specialized towards single views of regular objects, such as objects containing parallel structures.

In [109] Shapiro *et al.* define the epipolar geometry between pairs of *affine cameras* and describe a robust algorithm to compute its special fundamental matrix. Least squares formulations are employed in a noise-resistant fashion. A statistical noise model is employed to estimate the uncertainty of the computed relative pose of the cameras.

In [140] Zisserman *et al.* describe a technique to determine affine and metric calibration for a stereo rig that does not involve calibration objects but a single, general motion of the whole rig with fixed parameters. The internal camera parameters and the relative orientation of the cameras are retrieved robustly and automatically (see also [34]). This approach does not require solving the non-linear Kruppa equations [41, 70].

Related to the previous work is the paper by Zhang *et al.* [139]. This describes an algorithm for the self-calibration of a stereo rig and metric scene reconstruction using, again, the motion of the entire stereo rig, but in this case a simplified camera model is used (the principal point is known). Because of the exploitation of information redundancy in the stereo system, this approach yields to a more robust calibration result than only considering a single camera.

In [40] Faugeras *et al.* investigate a method to upgrade from projective to affine and finally metric structure from pairs of images by making use of scene constraints such as parallelism, orthogonality and known ratios of line segments. This work applies well to reconstructing architectural and aerial views from long sequences. A similar approach is used by Liebowitz *et al.* in [77] which make extensive use of *circular points* to elegantly upgrade from affine to metric structure.

2.3 Using three or more views

Two views suffice to reconstruct a scene, but adding one more image, taken from a third point of view, can constrain the reconstruction problem more, therefore reducing the uncertainty in the estimated structure. This is particularly true if a line matching process is used rather than a point matching one (line matching is not possible with only two views). Furthermore, the use of three or more views allows a check to be made on the consistency of the features matched using the first two views.

Faugeras was one of the first to investigate the problem of using more than two images. In [41] Faugeras *et al.* present a method for self-calibrating a camera using triplets of images from a sequence. The authors demonstrate that it is possible to calibrate a camera just by pointing it at the environment, selecting points of interest and then tracking them in the image sequence as the camera moves (the internal parameters of the camera are assumed to be constant throughout the sequence). The motion of the camera does not need to be known. The solution is obtained by solving the Kruppa equations.

Work on reconstruction from multiple views has been conducted by Avidan and Shashua in [4] and by Shashua in [112, 113], where the geometry involved in a three-dimensional reconstruction task from 2D images is analyzed in a very systematic and algebraic way. The concept of *trifocal tensor*, which encapsulates the geometric relationship between three views, is used [56, 116]. This theory is also found useful for solving the "rendering" problem, i.e. from the knowledge of two images, predicting what a third one will be if the camera is placed in a new, given position [43].

The problem of calibrating the camera and estimating the Euclidean structure of the scene using three or more uncalibrated images has been investigated also by Armstrong *et al.* in [2, 3]. The authors, using only point matches between images, compute the internal camera parameters and the Euclidean structure of the viewed scene. The camera undergoes a planar motion. This method presents the advantage that it can be applied in active vision tasks, the Euclidean reconstruction can be obtained during normal operation with initially uncalibrated cameras; i.e. the cameras do not need to be calibrated off-line. The trifocal tensor is used here too.

In [55] another approach for camera calibration using three or more images from a rotating camera is presented by Hartley. This method does not require the knowledge of the camera orientation and position and does not involve the epipolar geometry. In fact the images are taken from the same point of view and therefore the epipolar geometry cannot be used. The calibration process is based on image correspondences only and the internal parameters are assumed constant.

In [94] Pollefeys *et al.* investigate the problem of self-calibration and reconstruction from sequences with varying camera parameters. By employing a linear approximation followed by a non-linear refinement step, they achieve an accurate three-dimensional metric structure of the viewed scene while computing the internal parameters in each frame.

More recently, many commercial products have been developed which reconstruct three-dimensional architectural scenes from images. In particular *Facade* [31] has been one of the most successful systems for modelling and rendering architectural buildings from photographs. It consists of a hybrid geometric- and image-based approach. Realistic rendering is obtained by using "view-dependent texture mapping". Unfortunately the model building process involves considerable time and effort for the user.

Canoma[1] is a fast, interactive system to create three-dimensional scenes from single or multiple images which makes extensive use of three-dimensional templates. Another example is provided by *PhotoModeler*[2] where a three-dimensional model of the viewed scene is computed from the knowledge of some control points.

An easy-to-use system is *PhotoBuilder*[3] developed by Cipolla *et al.* [16]. The user interactively selects corresponding points on two or more photographs. Then the system, making use of parallelism and orthogonality constraints, computes the intrinsic and extrinsic parameters of the cameras and thus recovers metric models of architectural buildings.

2.4 Partial calibration

Not much work has been reported in the literature about metric reconstruction from only partial calibration. A partially calibrated scene can provide only partial metric information.

In [96] Proesmans *et al.* use minimal information about the scene (vanishing points and known vertical reference heights) to compute the heights of people from forensic photographs. The authors make extensive use of the properties of cross-ratios and address the possibility of employing geometric information extracted from shadows. They also propose the Cascaded Hough Transform algorithm for the automatic computation of vanishing points in single views.

In [130] Weinshall *et al.* make use of partial scene calibration, inferred directly from multiple images to obtain ordinal distances. A stratified approach is presented to upgrade the reconstruction to affine and finally Euclidean, using the available scene information. This technique employs two views and a plane-plus-parallax approach.

The ideas presented in all the above work are explored and generalized in the novel, single- and multi-view reconstruction techniques presented in this book.

---

1 `http://www.metacreations.com`
2 `http://www.photomodeler.com`
3 `http://www-svr.eng.cam.ac.uk/photobuilder/download.html`

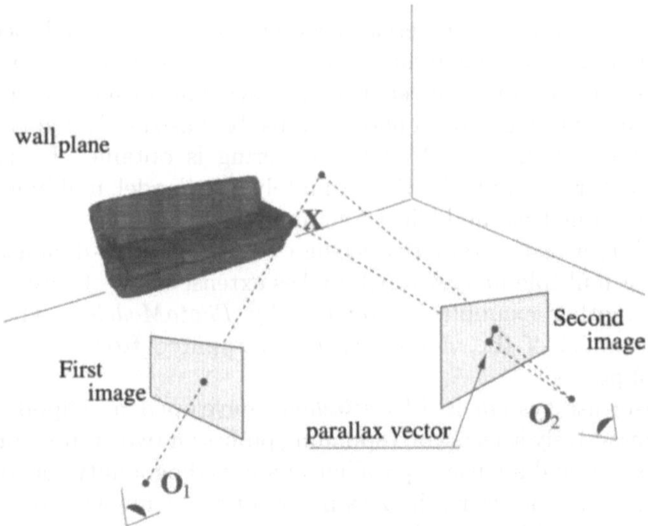

Fig. 2.3. Parallax geometry: the parallax vector detected on the second view is related to the distance of the three-dimensional point **X** from the wall plane.

2.5 Parallax-based approaches

Some other reconstruction methods in the literature make use of the concept of *parallax*.

One view only constrains each image point to lie on its projection ray but, in order to obtain the depth of a three-dimensional point with respect to a reference plane (or a generic surface), a second view is in general necessary. The parallax vector seen in the second image (see Fig. 2.3) is, in fact, a function of the depth of the point from the reference plane.

The parallax approach has been used by Cipolla in [15] and by Cipolla and Giblin in [17] as a way to estimate robustly the curvature of surfaces. They show that, in order to recover depth from motion, usually the rotational velocity must be known accurately. But this dependence can be removed if, instead of using raw motion, the difference of the image motions of points (parallax) is used. The deletion of the rotational component of the motion field from the process of depth estimation can considerably improve the accuracy of the reconstruction.

Cipolla et al. employed *affine parallax* in [74] for epipolar geometry computation using the parallax field; and in [75] for the estimation of the ego-motion of a mobile robot. Cipolla et al. also need this technique in [18] for three-dimensional structure estimation. Such work establishes once more the validity and effectiveness of this approach.

In his thesis [115] Sinclair discusses a method of analyzing sequences of images and identifying points which do not lie on the same planar surface. This work makes extensive use of planar projectivities (homographies). The parallax effect shown by points not lying on the same plane is employed for obstacle detection purposes. An analysis of the point localization error is developed to assess the limit to which noisy points can be considered to belong to the same plane.

In [105] Sawhney presents a formulation for three-dimensional motion and structure recovery that employs the parallax concept with respect to an arbitrary plane in the environment (real or virtual dominant plane). He shows that, if an image coordinate system is warped using a plane-to-plane homography with respect to a reference view, then the residual image motion depends only on the epipoles and has a simple relation to the three-dimensional structure.

Directly related to the work of Sawhney is that of Kumar *et al.* [71] where the authors show that the magnitude of the parallax vector associated with a point is directly proportional to its distance from a reference plane and inversely proportional to its depth from the camera. Furthermore in [72] they show how scenes can be represented implicitly as a collection of images. In this work a parallax-based approach is very effective for independent motion detection, pose estimation and construction of mosaic representations.

The geometry of the reconstruction problem has been discussed in depth and used by several researchers. In fact, a good understanding of the geometric properties of the acquisition system can be very useful for computing invariant quantities which increase the robustness and accuracy of the process [38, 58, 89].

The idea of using parallax to compute geometric invariants has been investigated by Irani and Anandan in [61] and Irani et al. in [62]. They present an analytical relationship between the image motion of pairs of points and several invariant quantities which can be derived from that relationship. This is shown to be useful for scene structure recovery from more than two images as well as for the detection of moving objects and the synthesis of new camera views.

In this book a ***plane-plus-parallax*** approach is developed in Chapter 6 for computing scene structure and camera location, and for recovering motion and structure constraints.

2.6 Investigation of accuracy

The above sections have presented an overview of the relevant work done in reconstructing three-dimensional models of a scene from the analysis of planar images.

But metric information is no use without an appropriate estimation of the uncertainties on the computed distances. In fact, visual metrology must be treated as a precision engineering task; i.e. in a reliable, efficient and accurate way [38].

Faugeras [38] and Kanatani [64] have presented general analytical methods for error analysis propagation based on analysis of covariances and how these are affected by the chain of computations. Formulae to compute uncertainties of matrices, vectors and points are discussed and the powerful ***implicit function theorem*** introduced.

The problem of error estimation is discussed in [132] by Weng *et al.* In this work the authors present an algorithm to estimate motion parameters and three-dimensional scene structure from two perspective images. They also investigate how the noise corrupting the perspective projections of feature points affects the computed motion parameters. This is achieved by means of a first order *matrix perturbation theory* [50, 120].

In [107] Shapiro and Brady analyze an outlier rejection algorithm based on a *principal component regression* framework. An error estimation method is conducted using the matrix perturbation theory and employed to identify and reject outliers. The authors also provide an error analysis for the optimal solution once the outliers have been removed. The validity of the proposed scheme is demonstrated by applying it to a structure-from-motion (SFM) problem where only affine cameras are considered.

Csurka *et al.* in [29] and Zhang in [135] compute the uncertainty related to the epipolar geometry between two cameras. They use the eight-point algorithm to compute the fundamental matrix and then estimate the associated covariance matrix. Two approaches are used to compute the uncertainty in the fundamental matrix: the first one is statistical and therefore as accurate as required but time consuming; the second is analytical and much simpler.

The first order estimation theory is analyzed and used extensively in this book, including examples of propagating uncertainty through explicit transformations, transformations defined by implicit functions and even implicit functions with constraints. Working examples are presented showing how the theory developed can be applied to well-known reconstruction problems.

In [73] Kumar and Hanson, using the parallax approach try to estimate robustly the camera location and orientation (*camera pose*) from a matched set of three-dimensional model and two-dimensional landmark features. Robust techniques for pose determination are developed to handle data contaminated by fewer than 50% outliers. The sensitivity of pose determination to incorrect estimates of camera parameters is analyzed.

Mohr *et al.* in [86] discuss methods for accurate reconstruction from multiple images. The cameras internal parameters are unknown. The projective reconstruction is upgraded to Euclidean by identifying the plane at infinity and then imposing metric constraints. A simple theory for analysing the accuracy of the reconstructed scene is presented. The need for a sub-pixel image point detection is shown to be necessary to obtain accurate three-dimensional structure.

2.7 Projective geometry and history of art

This work also presents contributions in understanding visual arts. Since art history has always been perceived as a non-scientific area, little attempt has been made to analyze paintings in a mathematical way and, possibly, reconstruct the represented scenes.

In [67] Kemp investigates Western art from a scientific point of view. In particular he describes the theory of linear perspective and its usage in the Renaissance period. The application of perspective rules to paintings is described in his work by analyzing (by construction) shapes and location of painted objects. It is quite interesting to approach the same problems from an algebraic, projective geometry-based, point of view. The author also investigates the theory of light and colour up to the French movements of the nineteenth century.

Kemp's book demonstrates that art history would benefit from the powerful tools offered by computer vision. On the other hand, vision researchers, in an attentive analysis of the relevant paintings, would find new explanations and a better understanding of the perception of three-dimensional space.

In [47] Field *et al.* meticulously describe the geometric technique they have employed to analyze the perspective in Masaccio's fresco *La Trinità* (*La Trinità con la Vergine e San Giovanni*, fresco, 1427/28, Santa Maria Novella, Firenze, Italy; see Fig. 7.2**a** and colour plate 2). Masaccio's Florentine masterpiece is recognized by many as the first painting where Brunelleschi's perspective rules have been applied. Probably this is the first perspective image in history.

The same fresco has also been investigated by De Mey [30]. By analyzing the construction lines (visible in X rays beneath the surface of the fresco), he computed a three-dimensional CAD model of the painted chapel and created animations. The location of the vantage point is also computed.

Much interest has been generated in paintings by Piero della Francesca (1416–1492), a very skilled artist and mathematician [32] from the Renaissance period. In particular, his ability with perspective drawings and his "religious" care of details have made his masterpiece *Flagellazione* (approx. 1453, Galleria Nazionale delle Marche, Urbino, Italia) one of the most studied perspective paintings [45].

Judith Field, in her recent book [46], presents a history of the evolution of mathematics and optics, and their usage in visual arts from the middle ages to the seventeenth century. She also describes how, after the seventeenth century, a number of mathematicians (e.g. Desargues, Pascal and Taylor) become increasingly interested in linear perspective [122], thus setting the foundations of modern *projective geometry.*

In this book, projective geometry-based methods for computing the structure of painted scenes and creating convincing animations are presented in Chapter 5. Results and examples are also shown in Chapter 7.

3. Background geometry and notation

Projective geometry [7, 38, 106] provides this book with the basic mathematical background, on top of which an effective and robust metrology framework is developed. This chapter presents the notation conventions and specific details of projective geometry which will be employed in the later chapters.

Notation.
This book employs quite standard notation conventions:

- three-dimensional points in general positions are denoted by upper case bold symbols (e.g. $\mathbf{X}$);
- image positions and vectors are denoted by lower case bold symbols (e.g. $\mathbf{x}$, $\boldsymbol{\mu}_p$);
- scalars are denoted by italic face symbols (e.g. d, Z_p);
- matrices are denoted by typewriter style capitals (e.g. $\mathtt{P}$, $\mathtt{H}$);
- the line through two points $\mathbf{x}_1$ and $\mathbf{x}_2$ is denoted by $< \mathbf{x}_1\mathbf{x}_2 >$.

When necessary, further notation conventions will be described in each chapter.

3.1 Camera models and perspective mappings

The image formation process must be modelled in a rigorous mathematical way. This section describes the camera models and the projective transformations which are relevant in the remainder of the book.

3.1.1 Pinhole camera model

The most general linear camera model is the well-known central projection (*pinhole camera*).

Description.
A three-dimensional point in space is projected onto the image plane by means of straight visual rays. The corresponding image point is the intersection of the image plane with the visual ray connecting the optical centre and the three-dimensional point (cf. Leonardo's "Perspectograph" in Fig. 3.1 and the schematic pinhole camera model in Fig. 3.2).

Fig. 3.1. "Perspectograph" (detail), by Leonardo da Vinci (1452–1519), Codex Atlanticus c.5r. Property of the Ambrosian Library, Milano.
"The things approach the point of the eye in pyramids, and these pyramids are intersected on the glass plane."

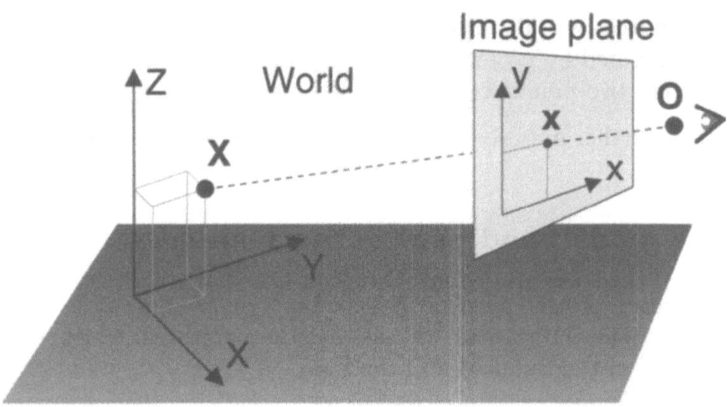

Fig. 3.2. Pinhole camera model: a point **X** in the three-dimensional space is imaged as **x**. Euclidean coordinates X, Y, Z and x, y are used for the world and image reference systems, respectively. **O** is the centre of projection, the viewer.

Algebraic parametrization.
The projection of a world point **X** onto the image point **x** (Fig. 3.2) is described by the following equation:

$$\mathbf{x} = \mathsf{P}\mathbf{X} \tag{3.1}$$

where P is the projection matrix, a 3×4 homogeneous matrix, and "=" is equality up to scale. The world and image points are represented by homogeneous

Fig. 3.3. An image of a planar surface induced by a homography: the homography relates points on the world plane to points on the image and viceversa.

vectors as $\mathbf{X} = (X, Y, Z, W)^\top$ and $\mathbf{x} = (x, y, w)^\top$. The scale of the matrix does not affect the equation, so only the eleven degrees of freedom corresponding to the ratio of the matrix elements are significant.

The camera model is completely specified once the matrix P is determined. The matrix can be computed from the relative positioning of the world points and camera centre, and from the camera internal parameters; however, it can also be computed directly from image-to-world point correspondences.

3.1.2 Planar homography

An interesting specialization of the general central projection described above is a plane-to-plane projection: a 2D-2D projective mapping.

Description.
The camera model for perspective images of planes, mapping points on a world plane to points on the image plane (and viceversa), is well known [106]. Points on a plane are mapped to points on another plane by a ***plane-to-plane homography***, also known as a plane projective transformation. It is a bijective (thus invertible) mapping induced by the star of rays centred in the camera centre (centre of projection). Planar homographies arise, for instance, when a world planar surface is imaged (see Fig. 3.3).

Algebraic parametrization.
A homography is described by a 3×3 non-singular matrix. Figure 3.4 shows the imaging process. Under perspective projection, corresponding points are related by [88, 106]:

$$\mathbf{x} = \mathtt{H}_i \mathbf{X} \tag{3.2}$$

where $\mathtt{H}_i$ is the 3×3 homogeneous matrix which describes the homography, and "=" is equality up to scale. The world and image points are represented by homogeneous 3-vectors as $\mathbf{X} = (X, Y, W)^\top$ and $\mathbf{x} = (x, y, w)^\top$ respectively.

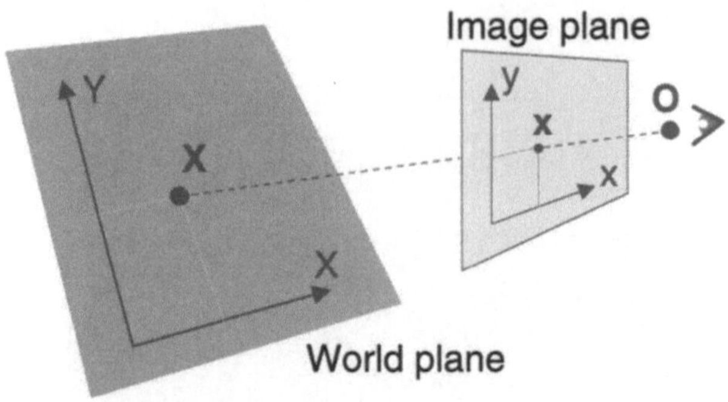

Fig. 3.4. Plane-to-plane camera model: a point **X** on the world plane is imaged as **x**. Euclidean coordinates X, Y and x, y are used for the world and image planes, respectively. **O** is the viewer's position.

The scale of the matrix does not affect the equation, thus only the eight degrees of freedom corresponding to the ratio of the matrix elements are significant.

Since we are interested in recovering world quantities from images, and the homography is an invertible transformation, in the following Chapters (3.2) will be written as:

$$\mathbf{X} = \mathtt{H}\mathbf{x} \tag{3.3}$$

with $\mathtt{H} = \mathtt{H}_i^{-1}$ the homography mapping image points into world points.

The camera model is completely specified once the matrix $\mathtt{H}$ is determined. Here too the matrix can be computed from the relative positioning of the two planes and camera centre, and the camera internal parameters; however, it can also be computed directly from image-to-world correspondences. In particular, it can be shown that at least four world-to-image feature (point or line) correspondences (no three points collinear or no three lines concurrent) suffice to define the homography. The relative geometric position of the world features (referred to as *computation points/lines*) must be known. This computation is described in section 4.1.

Inter-image homography.
A planar surface viewed from two different viewpoints induces a homography between the two images. Points on the world plane can be transferred from one image to the other (Fig. 3.5) by means of a homography mapping. This situation is considered in Chapter 6 where the parallax effect of points off the distinguished plane is analyzed.

3.1.3 Planar homology

In this section another interesting projective transformation is described: a planar homology. This is used extensively in the following chapters.

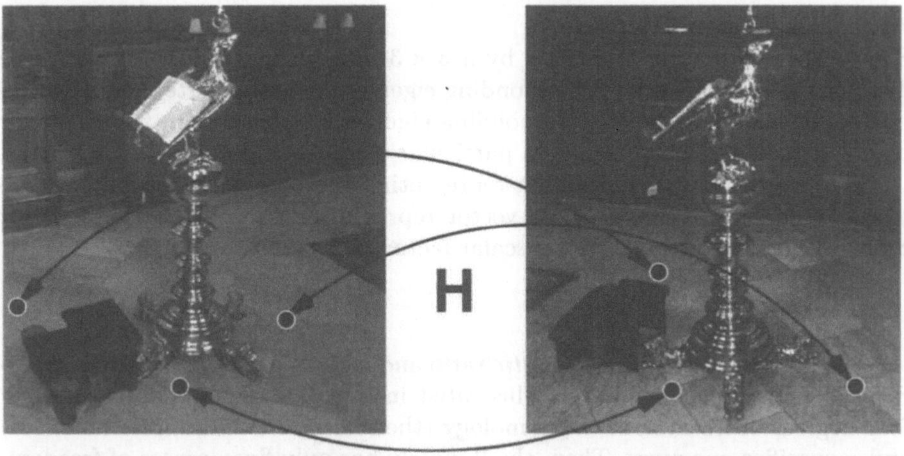

Fig. 3.5. Inter-image homography: the floor of a chapel viewed in both images induces a homography. Points can be mapped from one image to the other.

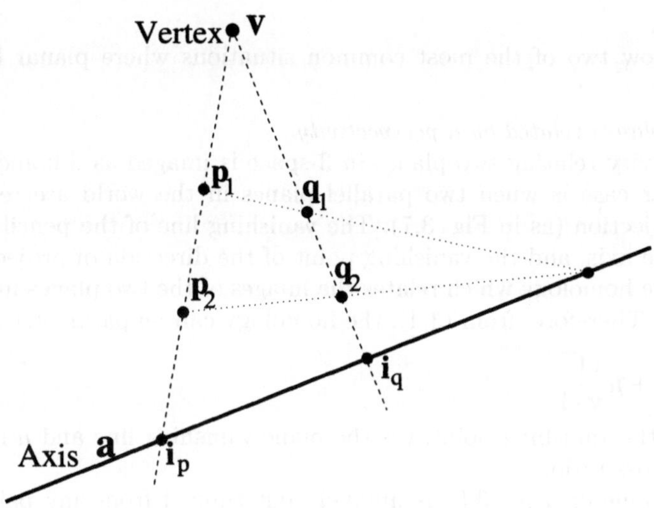

Fig. 3.6. A planar homology is defined by a vertex (the point $\mathbf{v}$), an axis (the line $\mathbf{a}$) and a characteristic ratio: its characteristic invariant ratio is given by the cross-ratio $< \mathbf{v}, \mathbf{p}_1, \mathbf{p}_2, \mathbf{i}_p >$ where $\mathbf{p}_1$ and $\mathbf{p}_2$ are *any* pair of points mapped by the homology and $\mathbf{i}_p$ is the intersection of the line through $\mathbf{p}_1$ and $\mathbf{p}_2$ and the axis. The point $\mathbf{p}_1$ is projected onto the point $\mathbf{p}_2$ under the homology, and similarly for $\mathbf{q}_1$ and $\mathbf{q}_2$. Note also that $< \mathbf{v}, \mathbf{p}_1, \mathbf{p}_2, \mathbf{i}_p > = < \mathbf{v}, \mathbf{q}_1, \mathbf{q}_2, \mathbf{i}_q >$.

Description.
A *planar homology* is a plane-to-plane projective transformation and a specialization of the homography. It is characterized by a line of fixed points called the *axis*, and a distinct fixed point not on the axis known as the *vertex* (Fig. 3.6). Planar homologies arise in several imaging situations, for instance when different light sources cast shadows of an object onto the same plane.

Algebraic parametrization.
Such a transformation is defined by a 3×3 non-singular matrix $\mathtt{H}$, with one distinct eigenvalue whose corresponding eigenvector is the vertex, and two repeated eigenvalues whose corresponding eigenvectors span the axis. A planar homology can be interpreted as a particular planar homography.

The projective transformation representing the homology can be parametrized directly in terms of the 3-vector representing the axis $\mathbf{a}$, the 3-vector representing the vertex $\mathbf{v}$, and a scalar factor μ as [129]:

$$\mathtt{H} = \mathtt{I} + \mu \frac{\mathbf{v}\mathbf{a}^\top}{\mathbf{v} \cdot \mathbf{a}} \tag{3.4}$$

The factor μ is the *characteristic* ratio and it can be computed as the cross-ratio of four aligned points as illustrated in Fig. 3.6. Thus $\mathbf{v}$ and $\mathbf{a}$ specify four degrees of freedom of the homology (the scales of $\mathbf{v}$ and $\mathbf{a}$ have no effect) and μ specifies one more. Then, the $\mathtt{H}$ matrix has only five degrees of freedom, (while a homography has eight dof) and thus it can be determined by 2.5 point correspondences.

Examples.
Here we show two of the most common situations where planar homologies arise.

Images of planes related by a perspectivity.
A perspectivity relating two planes in 3-space is imaged as a homology [128]. A particular case is when two parallel planes in the world are related by a parallel projection (as in Fig. 3.7). The vanishing line of the pencil of parallel planes is the axis, and the vanishing point of the direction of projection is the vertex of the homology which relates the images of the two planes in the pencil (Fig. 3.7**b**). Therefore, from (3.4), the homology can be parametrized as:

$$\mathtt{H} = \mathtt{I} + \mu \frac{\mathbf{v}\mathbf{l}^\top}{\mathbf{v} \cdot \mathbf{l}} \tag{3.5}$$

where $\mathbf{v}$ is the vanishing point, $\mathbf{l}$ is the plane vanishing line and μ is the characteristic cross-ratio.

The parameter μ in (3.5) is uniquely determined from any pair of image points which correspond between the two planes. Once the matrix $\mathtt{H}$ is computed, each point on a plane can be transferred into the corresponding point on the parallel plane. An example of this *homology mapping* is shown in Fig. 3.7**c**.

This configuration is extremely useful in Chapter 5 to compute affine measurements from single views. Notice that homologies arise in the equivalent case, when a sequence of images of one plane is taken by a purely translating camera; this is described in section 4.4. A homology also relates images of points on a plane in different views taken by a camera purely rotating about the axis orthogonal to the plane.

Desargues configuration.
Two triangles on a plane related such that the lines joining their corresponding vertices are concurrent are said to be in a *Desargues configuration*. Desargues' theorem states that the intersections of their corresponding sides are

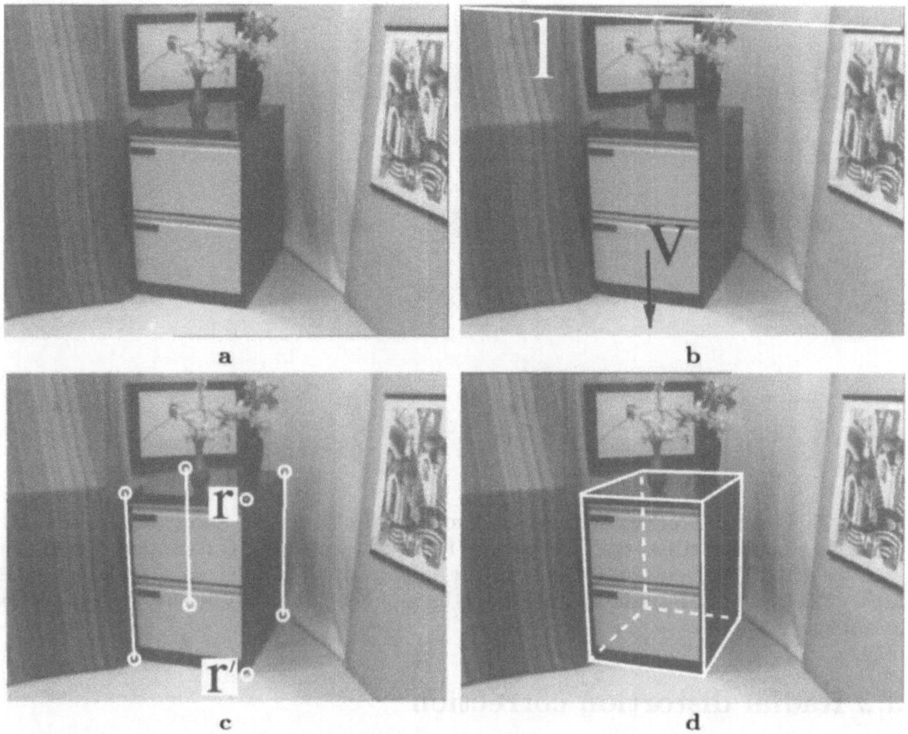

Fig. 3.7. Homology mapping of points from one plane to a parallel one, the two planes being related in the world by a parallel projection. **a** Original image, the floor and the top of the filing cabinet are parallel planes. **b** Their common vanishing line (axis of the homology, shown in white) has been computed by intersecting two sets of horizontal edges. The vertical vanishing point (vertex of the homology) has been computed by intersecting vertical edges. **c** Two reference corresponding points $\mathbf{r}$ and $\mathbf{r}'$ are selected on the two planes and the characteristic ratio of the homology computed. Three corners of the top plane of the cabinet have been selected and their corresponding points on the floor computed by the homology. Occluded corners have been retrieved too. **d** The wire-frame model shows the structure of the cabinet; occluded sides are dashed.

collinear [106] (see Fig. 3.8). Such triangles are related by a planar homology, with the common point of intersection being the vertex of the transformation, and the axis being the line containing the intersections of corresponding sides. Conversely, any triplet of points in correspondence under a homology must be in a Desargues configuration.

This configuration arises in real images, for example, when shadows of objects, originating from several light sources, are cast onto the same plane (see section 5.6). In section 6.2 the link between homologies and the structure and motion is derived in a plane-plus-parallax framework.

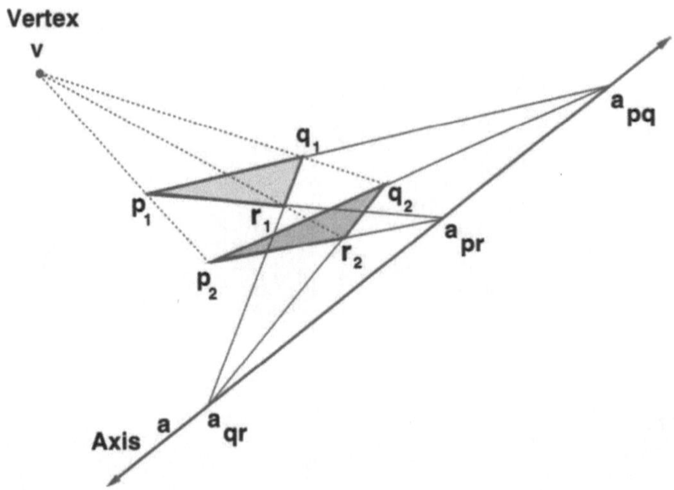

Fig. 3.8. Desargues configuration: the two triangles $\mathbf{p}_1, \mathbf{q}_1, \mathbf{r}_1$ and $\mathbf{p}_2, \mathbf{q}_2, \mathbf{r}_2$ are in a Desargues configuration; the lines joining their corresponding vertices are concurrent and, as stated by Desargues theorem, the intersections of their corresponding sides are collinear. The two triangles are related by a planar homology where $\mathbf{v}$ is the vertex and $\mathbf{a}$ the axis.

3.2 Radial distortion correction

A prerequisite of the theory treated in this book is that the camera behaves according to the pinhole model.

The problem.
This requirement is violated by cheap wide-angle lenses, such as those used in security systems. Those cameras, in fact, tend to distort the image, especially near its boundaries. In such cases the grossest distortions from the pinhole model are usually radial.

One of the main features of a pinhole camera is that it maps straight lines in the world into straight lines in the image. A radially distorted image, in contrast, shows straight scene lines mapped into curves. This may affect considerably any metrology algorithm. A correction step is, therefore, necessary before any metrology process may be performed.

Image correction.
Several possible methods have been investigated [65, 118] to correct such a distortion. A simple one has been proposed by Devernay and Faugeras [33], where only one image of the scene is necessary, and the radial distortion model is computed from the deformation of images of world straight edges.

Devernay's algorithm has been implemented here employing a slightly more complex radial distortion model to increase the correction accuracy:

$$\mathbf{x}_c = \mathbf{c} + f(r_d) * (\mathbf{x}_d - \mathbf{c}) \tag{3.6}$$

with:

$$f(r_d) = 1 + k_1 r_d + k_2 r_d^2 + k_3 r_d^3 + k_4 r_d^4, \quad r_d = d(\mathbf{x}_d, \mathbf{c})$$

where $\mathbf{x}_d$ is a point in the distorted image and $\mathbf{x}_c$ the corresponding point in the corrected image. $\mathbf{c}$ is the centre of distortion. $f(r_d)$ is the correction factor; it depends on the radius r_d ($d(\mathbf{a}, \mathbf{b})$ is the distance between the points $\mathbf{a}$ and $\mathbf{b}$) and the correction coefficients k_i. Therefore, if the values of the k_i coefficients are known, then (3.6) can be applied and the distorted image corrected.

In the model, account is taken of the fact that the centre of distortion might not coincide with the principal point of the image. Point coordinates are taken with respect to the centre of the image and normalized. Unit aspect-ratio and square pixels have also been assumed. The radial correction algorithm simply computes the best distortion parameters ($\mathbf{c}$, k_i $\forall i = 1 \cdots 4$) which transform the selected curved edges into straight ones.

An example.
Figure 3.9**a** shows an image captured from a cheap security-type camera which exhibited radial distortion (note how straight edges in the scene appear curved in the image).

Edges have been extracted automatically by a Canny edge detector [10]. A set of edges assumed to be straight in the scene has been selected (Fig. 3.9**b**) and from those the distortion parameters computed and the image corrected accordingly (Fig. 3.9**c**). Note that now images of straight edges in the world are straight (for example, the left edge of the filing cabinet). Such an image is now perspective, and metrology algorithms can safely be performed.

3.3 Vanishing points and vanishing lines

Vanishing points and vanishing lines are extremely powerful geometric cues; they convey a lot of information about the direction of lines and orientation of planes. These entities can be estimated directly from the images and *no* explicit knowledge of the relative geometry between the camera and the viewed scene is required [21, 78, 84, 114]. Often they lie outside the physical image (as in Fig. 3.7**b**), but this does not affect the computations.

Straight edges.
The first step towards the computation of vanishing points and lines is the accurate detection of straight edges of objects. Straight-line segments are detected in an image at sub-pixel accuracy by using a Canny edge detector [10] followed by edge linking, then segmentation of the edgel chain at high curvature points and finally straight-line fitting by orthogonal regression to the resulting chain segments (an example is shown in Fig. 3.10**b**).

Furthermore, lines which are the projection of a physical edge in the scene might appear broken in the image because of occlusions. A simple merging algorithm based on orthogonal regression has been implemented to merge several manually selected edges together. Merging aligned edges to create longer ones increases the accuracy of their location and orientation. An example is shown in Fig. 3.10**c**.

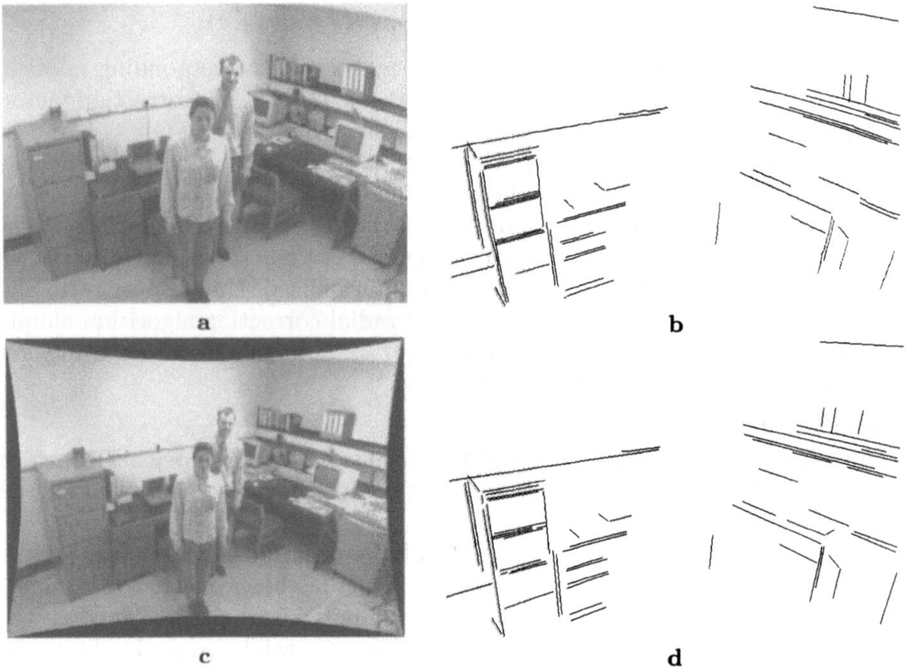

Fig. 3.9. Radial distortion correction: **a** original image showing radial distortion, note that straight edges in the scene appear curved in the image (e.g. the left edge of the filing cabinet); **b** lines corresponding to straight world edges have been selected in image (a) and used to compute the following distortion parameters: $k_1 = 0.092$, $k_2 = -0.007$, $k_3 = 0.053$, $k_4 = -0.012$ (the centre of distortion has been assumed in this case at the centre of the image); **c** corrected image; **d** edges from corrected image; images of straight scene edges are now straight.

Computing vanishing points.
Images of parallel world lines intersect each other in the same vanishing point. This is, therefore, defined by at least two such lines. However, if more than two lines are available, a *maximum likelihood estimate* (MLE) algorithm is employed to estimate the point [78] (see Fig. 3.11).

Computing vanishing lines.
Images of lines parallel to each other and to a plane intersect in points on the plane vanishing line. Therefore two sets of those lines with different directions are sufficient to define the plane vanishing line (see Fig. 3.12). A maximum likelihood estimation algorithm is employed if more than two orientations are available.

3.4 Uncertainty analysis

This section describes the basic ideas underlying the analysis of uncertainties treated in this book.

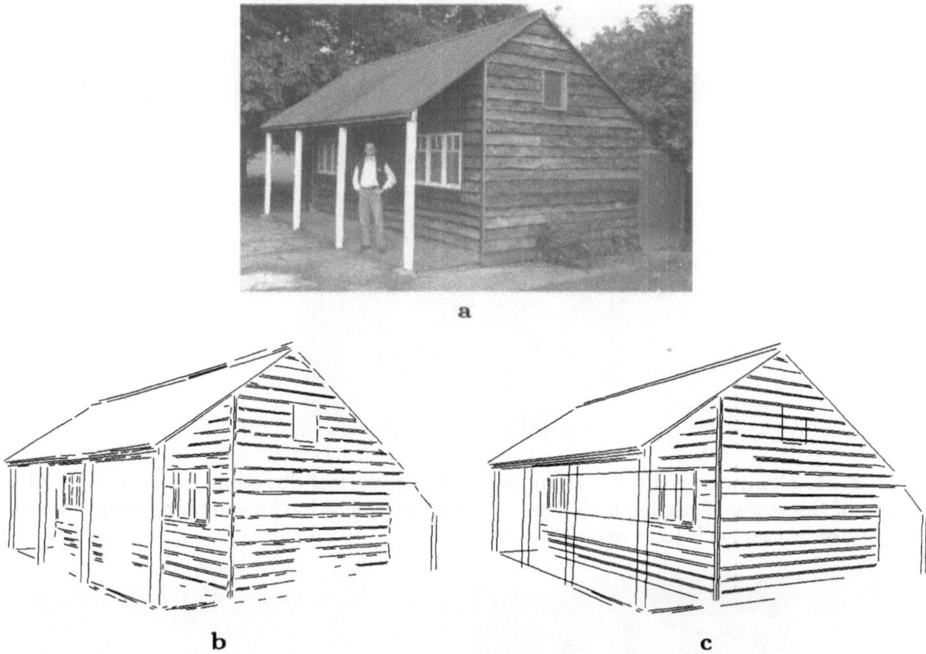

Fig. 3.10. Computing and merging straight edges: **a** original image; **b** some of the edges detected by the Canny edge detector. Straight lines have been fitted to them. **c** edges after the merging process – different segments belonging to the same edge in space are merged together.

The metrology algorithms described in the following chapters rely on applying non-linear transformations to image and world quantities (e.g. points and reference distances) to compute world measurements (e.g. distances between points and ratios of areas). Since the input quantities and the transformations are uncertain, the output measurements are uncertain too (Fig. 1.2). In order to determine how the uncertainty propagates from input to output of the computation chain and estimate the measurement accuracy, two methods are considered: a statistical one and an analytical one.

Uncertainties are modelled as *Gaussian* noise and described by the associated covariance matrix. For example, the uncertainty in the localization of a world point $\mathbf{X}$ is modelled as a bi-dimensional Gaussian (see Fig. 3.13) centred on the point itself. The uncertainty is described by a covariance matrix $\Lambda_{\mathbf{X}}$ and can be visualized also as an ellipse whose axes are given by the principal components of the related covariance matrix. It is not strictly necessary to have such idealized distributions but this has not been found to be a restriction in practice.

Statistical method.
This consists of an *iterative* algorithm which makes use of the *law of large numbers.*

Fig. 3.11. Computing the vanishing point of the vertical direction: images of parallel edges of the tower (Magdalen College, Oxford) intersect in the vanishing point (white circle); only two such lines are sufficient.

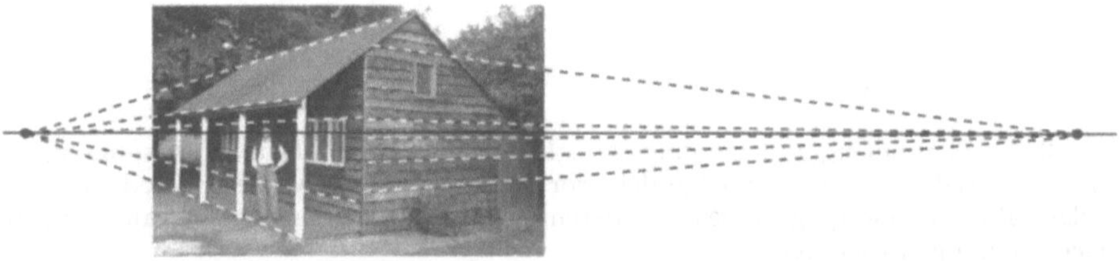

Fig. 3.12. Computing the vanishing line of the ground plane, the horizon. The vanishing line of the ground is shown in solid black. The planks on both sides of the shed define two sets of lines parallel to the ground (dashed); they intersect in points on the vanishing line. Two such points suffice to define the plane vanishing line.

By denoting with the input data by $\mathbf{x}$ and the geometric transformation that maps the input $\mathbf{x}$ into the final measurements $\mathbf{y}$ by $\mathbf{f}$, we can write:

$$\mathbf{y} = \mathbf{f}(\mathbf{x}) \tag{3.7}$$

At each iteration i, a random set of input data $\mathbf{x}_i$ is generated (according to its covariance $\Lambda_{\mathbf{x}}$); then a random output measurement, $\mathbf{y}_i$, is obtained from (3.7).

The process is repeated a large number of times N, thus obtaining a distribution of measurements around the true mean $\overline{\mathbf{y}}$. If N is large enough, then the mean measurement $\overline{\mathbf{y}}$ can be approximated by:

$$\overline{\mathbf{y}} \approx \frac{\sum_{i=1}^{N} \mathbf{y}_i}{N}$$

and its covariance by:

$$\Lambda_{\mathbf{y}} \approx \frac{\sum_{i=1}^{N} \left[(\mathbf{y}_i - \overline{\mathbf{y}})(\mathbf{y}_i - \overline{\mathbf{y}})^{\top}\right]}{N-1} \tag{3.8}$$

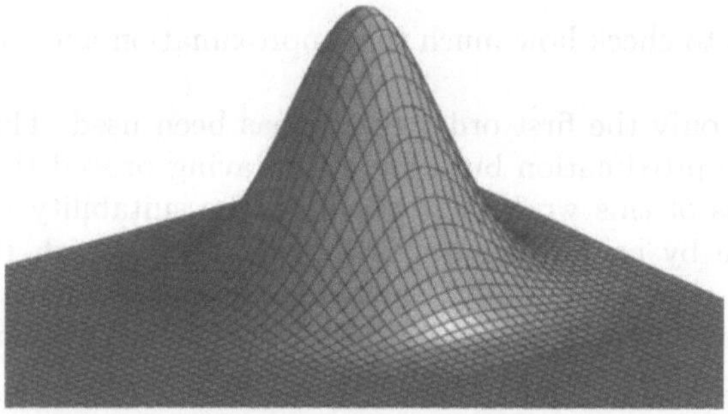

Fig. 3.13. Typical isotropic Gaussian density probability function: this is used to model the uncertainty in the localization of points on planes.

The covariance matrix $\boldsymbol{\Lambda}_{\mathbf{y}}$ encodes the uncertainty of the measurements. This method is very costly in terms of CPU time, and in fact it returns reliable results only after a very large number of iterations. It will be used in the following chapters to validate the much faster analytical method described below.

Analytical method.
The analytical method takes account of the fact that the measurement $\mathbf{y}$ is related to the input data $\mathbf{x}$ by an analytical function $\mathbf{f}$ (generally non-linear). This relationship is approximated with a linear one by means of a first order *Taylor series expansion*. By assuming noise only on the input data $\mathbf{x}$ and not on the transformation, (3.7) becomes [20, 38]:

$$\mathbf{f}(\overline{\mathbf{x}} + \Delta\mathbf{x}) = \mathbf{f}(\overline{\mathbf{x}}) + \nabla\mathbf{f}(\overline{\mathbf{x}})\Delta\mathbf{x} + \mathcal{O}(||\Delta\mathbf{x}||^2)$$

from which, ignoring second order terms, it is easy to compute the mean value of the output measurement:

$$\overline{\mathbf{y}} \approx E\left[\mathbf{f}(\overline{\mathbf{x}}) + \nabla\mathbf{f}(\overline{\mathbf{x}})\Delta\mathbf{x}\right] = E[\mathbf{f}(\overline{\mathbf{x}})] = \mathbf{f}(\overline{\mathbf{x}})$$

and consequently the covariance of the measurement $\boldsymbol{\Lambda}_{\mathbf{y}}$ is:

$$\begin{aligned}
\boldsymbol{\Lambda}_{\mathbf{y}} &= E([\mathbf{f}(\overline{\mathbf{x}} + \Delta\mathbf{x}) - \overline{\mathbf{y}}][\mathbf{f}(\overline{\mathbf{x}} + \Delta\mathbf{x}) - \overline{\mathbf{y}}]^\top) \\
&\approx E([\mathbf{f}(\overline{\mathbf{x}} + \Delta\mathbf{x}) - \mathbf{f}(\overline{\mathbf{x}})][\mathbf{f}(\overline{\mathbf{x}} + \Delta\mathbf{x}) - \mathbf{f}(\overline{\mathbf{x}})]^\top) \\
&\approx E(\nabla\mathbf{f}\Delta\mathbf{x}\Delta\mathbf{x}^\top\nabla\mathbf{f}^\top) \\
&= \nabla\mathbf{f}\boldsymbol{\Lambda}_{\mathbf{x}}\nabla\mathbf{f}^\top
\end{aligned}$$

In the case the relationship between input data and final measurement is described by an implicit function, then the covariance $\boldsymbol{\Lambda}_{\mathbf{y}}$ is obtained by making use of the *implicit function theorem* [20, 38]. Further analysis must be carried out in both the statistical and analytical methods in cases where the transformation $\mathbf{f}$ is also uncertain. When necessary these cases will be addressed in the relevant chapters.

The advantage in using such an analytical method is that it provides a non-iterative and therefore fast algorithm. The main disadvantage is that it introduces an approximation of the non-linear mapping function. Therefore a further analysis to check how much this approximation affects the final results is required.

In the past, only the first order theory has been used. This book will still use the linear approximation but only after having proved that it is sufficient for the purposes of this work. The check on the suitability of the first order analysis is done by comparing the first order results with the second order ones and with a number of Monte Carlo simulations, i.e. the statistical method described above is used as a test of fitness of the analytical approach.

Example.
In this example we compute the uncertainty in the localization of a point $\mathbf{x}'$ obtained from a homographic mapping of an uncertain point $\mathbf{x}$ on a different plane. The homography is described by the matrix $\mathtt{H}$ (such that $\mathbf{x}' = \mathtt{H}\mathbf{x}$) and is assumed to be known and exact. Uncertainty is assumed on the input point $\mathbf{x}$, its covariance being $\boldsymbol{\Lambda}_{\mathbf{x}}$.

Analytical method.
From the first order error analysis the uncertainty of the location of the point $\mathbf{x}'$ is:

$$\boldsymbol{\Lambda}_{\mathbf{x}'} = \mathtt{H}\boldsymbol{\Lambda}_{\mathbf{x}}\mathtt{H}^{\top} \tag{3.9}$$

Validation of first order analysis.
The statistical Monte Carlo method is now employed to assess the validity of (3.9). The test algorithm consists of the following three steps:

- 10 000 Gaussian random points centred about a mean value $\mathbf{x}$ with covariance $\boldsymbol{\Lambda}_{\mathbf{x}}$ are generated on the first plane;
- each point is transferred onto the second plane by the homography $\mathtt{H}$, thus creating a distribution of points;
- the covariance of this distribution is computed from (3.8) and compared to the one obtained from (3.9).

Figure 3.14 shows the results for synthetic data. Analytical and simulated covariance ellipses are almost overlapping, thus proving the validity of the first order approximation in this case.

Further details and examples on real data are presented in Chapter 4.

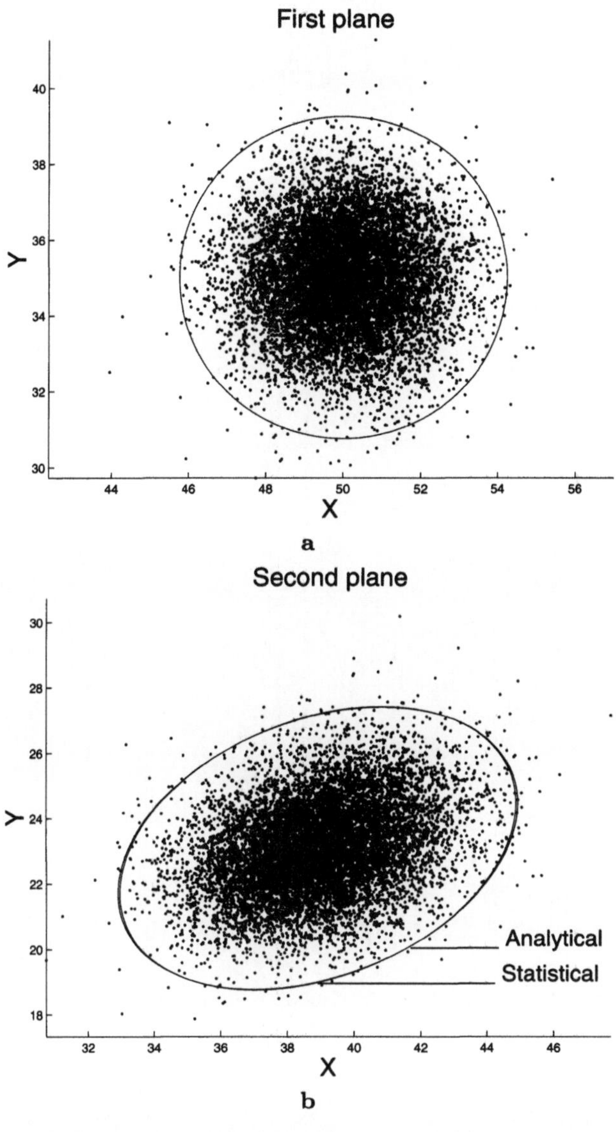

Fig. 3.14. Example of validation of first order error analysis by using a Monte Carlo test. **a** 10000 Gaussian points have been randomly generated about the mean position $\mathbf{x}$ with covariance $\Lambda_{\mathbf{x}}$ on the first plane. The corresponding 3-standard deviation ellipse is shown. **b** Distribution of the output points $\mathbf{x}'$ on the second plane. Both analytical and simulated 3-standard deviation ellipses are drawn. They overlap almost perfectly, confirming the validity of the analytical prediction.

4. Metrology on planes

This chapter develops an algorithm to measure distances on planar surfaces (see also [24]). The position, orientation and internal parameters of the camera are unknown. The imaging process is modelled as in section 3.1.2 as a homographic mapping between the world plane and the image plane.

Once the image-to-world homography matrix is determined, the back-projection of an image point to a point on the world plane is straightforward. The distance between two points on the world plane is then simply computed from the Euclidean distance between their back-projected images. The estimation of the homography matrix is described in section 4.1, and then that is used to obtain a number of world measurements.

Estimating the accuracy (or uncertainty) on planar distance measurements requires a careful treatment of the sources of error; not just the error in selecting the image points but also the error in the homography matrix itself. The latter arises from error in localization of the point correspondences from which the matrix is computed (computation points).

The uncertainty analysis developed here builds on and extends previous analysis of the uncertainty in relations estimated from homogeneous equations, for example homographies [99, 100] and epipolar geometry [29, 38]. It extends these results because it covers the cases both of where the matrix is exactly *and* where is over-determined by the world-image correspondences, and it is not adversely affected when the estimation matrix is near singular.

In this chapter, first order analysis is proved to be sufficient in typical imaging arrangement. It is also proved to be exact for the affine part of the homography and to be only an approximation for the non-linear part. An expression for the covariance of the estimated $\mathtt{H}$ matrix is derived in section 4.2.2. The correctness of the uncertainty predictions is tested both by Monte Carlo simulation (section 4.2.3) and by numerous experiments on real images (section 4.3). Examples on interior and architectural measurements are shown, together with examples on rectification of planes in images of real scenes and paintings from the Italian Renaissance. Finally, in section 4.4, further properties of the image rectification process are investigated.

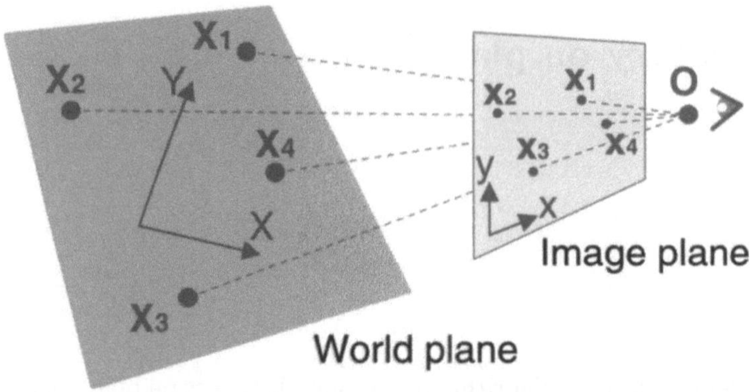

Fig. 4.1. Computing the plane-to-plane homography: at least four corresponding points (or lines) are necessary to determine the homography between two planes.

4.1 Estimating the homography

As seen in the introduction, the main problem is to estimate the homography between the viewed world plane and the image plane. This can be done directly from a set of known image-world correspondences (points or lines). In fact, if we consider only point correspondences, from (3.3) each image-to-world point correspondence provides two equations which are linear in the $\mathtt{H}$ matrix elements. They are:

$$\begin{aligned} h_{11}x + h_{12}y + h_{13} &= h_{31}xX + h_{32}yX + h_{33}X \\ h_{21}x + h_{22}y + h_{23} &= h_{31}xY + h_{32}yY + h_{33}Y \end{aligned}$$

For n correspondences we obtain a system of $2n$ equations in eight unknowns. If $n = 4$ (as in Fig. 4.1) then an exact solution is obtained. Otherwise, if $n > 4$, the matrix is over-determined, and $\mathtt{H}$ is estimated by a suitable minimization scheme.

The covariance of the estimated $\mathtt{H}$ matrix depends both on the errors in the position of the points used for its computation *and* the estimation method employed.

There are three standard methods for estimating $\mathtt{H}$:

i. Non-homogeneous linear solution. One of the nine matrix elements is given a fixed value, usually unity, and the resulting simultaneous equations for the other eight elements are then solved using a pseudo-inverse. This is the most commonly used method. It has the disadvantage that poor estimates are obtained if the chosen element should actually have the value zero.[1]

ii. Homogeneous solution. The solution is obtained using SVD (see section 4.1.1). This is the method used here and is explained in more detail below. It does not have the disadvantage of the non-homogeneous method but it minimizes an algebraic error which does not have a geometric meaning.

[1] This occurs, for instance, if we choose $h_{33} = 1$ and the origin of the image coordinate system lies on the vanishing line of the plane under consideration.

iii. Non-linear geometric solution. In this case the summed Euclidean distances between the measured and the mapped point is minimized, i.e. minimizing the following cost function:

$$\sum_{i=1}^{n}\left\{\left[\begin{pmatrix}X_i\\Y_i\end{pmatrix}-\frac{1}{(\mathtt{H}\mathbf{x}_i)_3}\begin{pmatrix}(\mathtt{H}\mathbf{x}_i)_1\\(\mathtt{H}\mathbf{x}_i)_2\end{pmatrix}\right]^2+\left[\begin{pmatrix}x_i\\y_i\end{pmatrix}-\frac{1}{(\mathtt{H}^{-1}\mathbf{X}_i)_3}\begin{pmatrix}(\mathtt{H}^{-1}\mathbf{X}_i)_1\\(\mathtt{H}^{-1}\mathbf{X}_i)_2\end{pmatrix}\right]^2\right\}$$

This returns the maximum likelihood estimation of the $\mathtt{H}$ matrix. This method has the advantage, over the above two algebraic methods, that the quantity minimized is meaningful and corresponds to the error involved in the measurement (similar minimizations are used to estimate the fundamental matrix and trifocal tensor [123, 137]). There is no closed form solution in this case and a numerical minimization scheme, such as Levenberg-Marquardt [95], is employed. Usually an initial solution is obtained by method (ii), and then "polished" using this method.

4.1.1 Homogeneous estimation method

Writing the $\mathtt{H}$ matrix as a 9-vector

$$\mathbf{h}=(h_{11},h_{12},h_{13},h_{21},h_{22},h_{23},h_{31},h_{32},h_{33})^\top$$

the homogeneous equation (3.3) for n points become

$$\mathtt{A}\mathbf{h}=\mathbf{0}$$

with $\mathtt{A}$ the $2n\times 9$ matrix:

$$\mathtt{A}=\begin{pmatrix}
x_1 & y_1 & 1 & 0 & 0 & 0 & -x_1X_1 & -y_1X_1 & -X_1\\
0 & 0 & 0 & x_1 & y_1 & 1 & -x_1Y_1 & -y_1Y_1 & -Y_1\\
x_2 & y_2 & 1 & 0 & 0 & 0 & -x_2X_2 & -y_2X_2 & -X_2\\
0 & 0 & 0 & x_2 & y_2 & 1 & -x_2Y_2 & -y_2Y_2 & -Y_2\\
\vdots & \vdots & \vdots & \vdots & \vdots & \vdots & \vdots & \vdots & \vdots\\
x_n & y_n & 1 & 0 & 0 & 0 & -x_nX_n & -y_nX_n & -X_n\\
0 & 0 & 0 & x_n & y_n & 1 & -x_nY_n & -y_nY_n & -Y_n
\end{pmatrix}$$

The problem of computing the $\mathbf{h}$ vector is now reduced to the constrained minimization of the following cost function:

$$C=\mathbf{h}^\top\mathtt{A}^\top\mathtt{A}\mathbf{h} \tag{4.1}$$

subject to the constraint that $||\mathbf{h}||=1$. The corresponding Lagrange function is:

$$\mathcal{L}=\mathbf{h}^\top\mathtt{A}^\top\mathtt{A}\mathbf{h}-\lambda(\mathbf{h}^\top\mathbf{h}-1) \tag{4.2}$$

Differentiating this with respect to $\mathbf{h}$ and setting these derivatives equal to zero we obtain:

$$\frac{\partial\mathcal{L}}{\partial\mathbf{h}}=2\mathtt{A}^\top\mathtt{A}\mathbf{h}-2\lambda\mathbf{h}=0$$

i.e.

$$\mathtt{A}^\top \mathtt{A}\mathbf{h} = \lambda \mathbf{h}$$

Therefore the solution $\mathbf{h}$ is an unit eigenvector of the matrix $\mathtt{A}^\top\mathtt{A}$ and $\lambda = \mathbf{h}^\top \mathtt{A}^\top \mathtt{A}\mathbf{h}$ is the corresponding eigenvalue.

In order to minimize the C function, only the eigenvector $\tilde{\mathbf{h}}$ corresponding to the minimum eigenvalue $\tilde{\lambda}$ should be considered. This eigenvector can be obtained directly from the Singular Value Decomposition (SVD) of $\mathtt{A}$. In the case of $n = 4$, $\mathbf{h}$ is the null-vector of $\mathtt{A}$ and the residuals are zero.

4.2 Uncertainty analysis

As shown in section 3.1.2, if $\mathbf{x}$ is a point on the image plane its back-projection onto the world plane is given by (3.3) (in homogeneous coordinates). However the input image point $\mathbf{x}$ is uncertain because of localization errors on the image. Furthermore, the homography is uncertain because it is obtained from $n \geq 4$ pairs of uncertain computation points (their location is known but uncertain). These two sources of error affect the localization of the output point $\mathbf{X}$ on the world plane and consequently the distance between two world points.

The goal here is to model the uncertainty in the localization of a world point and in the distance between two world points once the uncertainty in the localization of image points and computation pairs is known.

4.2.1 First and second order uncertainty analysis

In order to avoid unnecessarily complicated algebra, the comparison between first and second order analysis is developed for a line-to-line homography (see Fig. 4.2). The one-dimensional case illustrates all the ideas involved, and the algebraic expressions are easily interpreted. The generalization to 3×3 matrices is straightforward and does not provide any new insights here.

In the one-dimensional case, (3.3) reduces to:

$$\begin{pmatrix} X \\ 1 \end{pmatrix} = \mathtt{H}_{2\times2} \begin{pmatrix} x \\ 1 \end{pmatrix}$$

where $\mathtt{H}_{2\times2}$ is a 2×2 homography matrix. For the geometry shown in Fig. 4.2 the matrix is given by:

$$\mathtt{H}_{2\times2} = \begin{pmatrix} \alpha & t \\ \mu & 1 \end{pmatrix}$$

with parameters $\alpha = \frac{d}{f\cos^2(\omega)} - \frac{t}{f}\tan(\omega)$ and $\mu = -\frac{\tan(\omega)}{f}$. Under back-projection an image point x maps as:

$$x \rightarrow X = \frac{h_{11}x + h_{12}}{h_{21}x + h_{22}} = \frac{\alpha x + t}{\mu x + 1}$$

This non-linear mapping (in inhomogeneous coordinates) can be expanded in a Taylor series. Statistical moments of X, such as the variance, are then computed

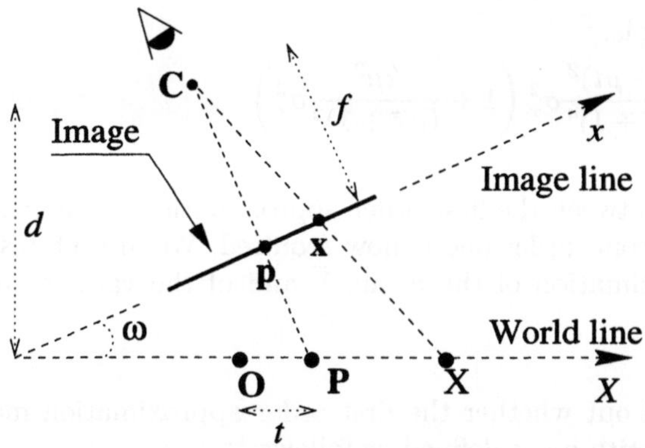

Fig. 4.2. One-dimensional camera model: the camera centre is at distance f (the focal length) from the image line. The ray at the principal point **p** is perpendicular to the image line, and intersects the world line at **P**, with world ordinate t. The angle between the world and image lines is ω.

in terms of the Taylor coefficients and the moments of x [22, 102]. It is assumed here that the homography is exact (no errors) and the measurement of the image test point x is subject to Gaussian noise with standard deviation σ_x. The Taylor series is developed about the point's mean position denoted as $\overline{x}$.

First order.
If the Taylor series is truncated to first order in $(x - \overline{x})$ then the mapping is linearized:

$$X \approx \frac{\alpha\overline{x} + t}{\mu\overline{x} + 1} + \frac{\alpha - \mu t}{(\mu\overline{x} + 1)^2}(x - \overline{x})$$

The mean of X is:

$$\overline{X} \approx \frac{\alpha\overline{x} + t}{\mu\overline{x} + 1} \tag{4.3}$$

since the mean is a linear operator. The variance of X can be shown to be:

$$\sigma_X^2 \approx \frac{(\alpha - \mu t)^2}{(\mu\overline{x} + 1)^4}\sigma_x^2 \tag{4.4}$$

Second order.
Usually only the first order approximation is used for error propagation [19, 29, 38, 99, 100]. Here the Taylor expansion is extended to second order so that the approximation involved in truncating to first order can be bounded. It can be shown that to second order in $(x - \overline{x})$:

$$X \approx \frac{\alpha\overline{x} + t}{\mu\overline{x} + 1} + \frac{\alpha - \mu t}{(\mu\overline{x} + 1)^2}(x - \overline{x}) - \frac{\mu(\alpha - \mu t)}{(\mu\overline{x} + 1)^3}(x - \overline{x})^2$$

The mean of X is:

$$\overline{X} \approx \frac{\alpha\overline{x} + t}{\mu\overline{x} + 1} - \frac{\mu(\alpha - \mu t)}{(\mu\overline{x} + 1)^3}\sigma^2{}_x \tag{4.5}$$

and its variance is:

$$\sigma_X^2 \approx \frac{(\alpha - \mu t)^2}{(\mu \overline{x} + 1)^4} \sigma_x^2 \left(1 + \frac{2\mu^2}{(\mu \overline{x} + 1)^2} \sigma_x^2\right) \tag{4.6}$$

Comparison.
A comparison between the first order approximation of the perspectivity function and the second order one is now required. We are interested in the error made in the estimation of the mean $\overline{X}$ and of the variance σ_X^2 of the world point X.

Bias on mean.
In order to find out whether the first order approximation mean is biased or not, a bias quantity b_X is defined as follows:

$$b_X = \overline{X}^{1^{st}} - \overline{X}^{2^{nd}} = \frac{\mu(\alpha - \mu t)}{(\mu \overline{x} + 1)^3} {\sigma^2}_x$$

Then the bias b_X is just the difference between the mean of X computed from the first order series truncation and that computed from the second order one. The bias has dimensions of length.

The Lagrange remainder of the Taylor series [121] provides an upper bound for the error on the mean if the series is truncated to first order instead of using the complete expansion. Using the first order truncation of the Taylor expansion with the Lagrange remainder we obtain:

$$\overline{X} = \frac{\alpha \overline{x} + t}{\mu \overline{x} + 1} - \frac{\mu(\alpha - \mu t)}{(\mu \xi + 1)^3} {\sigma^2}_x$$

with $\xi \in]x, \overline{x}[$. Therefore the upper bound of the error (absolute value) on the mean of X is:

$$Err_{\overline{X}} = \frac{\mu |\alpha - \mu t|}{(\mu \xi + 1)^3} {\sigma^2}_x$$

Thus a measure, M_b, can be defined as the bias in the mean between first and second order approximation using the ratio of the Lagrange upper bound of the mean $\overline{X}$ to the mean itself:

$$M_b = \frac{\mu |\alpha - \mu t| (\mu \overline{x} + 1)}{(\alpha \overline{x} + t)(\mu \xi + 1)^3} \sigma_x^2 \tag{4.7}$$

Error in variance.
Two new measures are defined here. They are used to assess the error in the standard deviation (or variance) in truncating to first order. The first, M_{v_1}, measures the ratio of the second order variance to the first order one. Comparing (4.4) and (4.6) this ratio is:

$$M_{v_1} = 2 \frac{\mu^2}{(\mu \overline{x} + 1)^2} \sigma_x^2 \tag{4.8}$$

The second measure, M_{v_2}, is obtained from the Lagrange remainder of the Taylor series; this provides an upper bound on the error if the series is truncated to first order instead of using the complete expansion. From Taylor's theorem the variance σ_X^2 is:

$$\sigma_X^2 = \frac{(\alpha - \mu t)^2}{(\mu\overline{x} + 1)^4}\sigma_x^2 + \frac{2\mu^2(\alpha - \mu t)^2}{(\mu\xi + 1)^6}\sigma_x^4$$

This is the sum of the first order term and the Lagrange remainder, which for some ξ such that $|\overline{x} - \xi| < |\overline{x} - x|$ is an exact equation, not an approximation. The measure M_{v_2} is then obtained as the ratio of this truncation error to the first order term:

$$M_{v_2} = 2\frac{\mu^2(\mu\overline{x} + 1)^4}{(\mu\xi + 1)^6}\sigma^2{}_x \tag{4.9}$$

and we compute the worst case of $\xi = \overline{x} - \sigma_x$ for this bound in the range $\overline{x} \pm \sigma_x$.

Typical results.
The significance of these measures is that they depend only on the elements of the homography matrix. Thus, once a matrix has been estimated the need for a second order approximation can be immediately assessed. All these measures are dimensionless and it makes sense to compare their values to unity.

In Fig. 4.3 a graph plot of the measures defined is given in order to visualize their behaviour with respect to variations of the angle ω. The three linearity measures are computed for the following values: $f = 8.5$ mm, $d = 5$ m, $t = 1$ m, $x = 50$ pixels, $\sigma_x = 1$ pixel, and ω varies from 0° to 85° (cf. Fig. 4.2). For angle values close to $70^o - 80^o$ the non-linearity of the projection function suddenly increases and the first error analysis is near the limit of its usefulness.

The same analysis is reported in the table below for discrete values of the angle ω:

ω	30°	40°	50°	60°	70°	80°
M_b	$5.9 \cdot 10^{-6}$	$1.1 \cdot 10^{-5}$	$1.9 \cdot 10^{-5}$	$3.9 \cdot 10^{-5}$	$9.5 \cdot 10^{-5}$	$4.2 \cdot 10^{-4}$
M_{v_1}	$1.5 \cdot 10^{-6}$	$3.3 \cdot 10^{-6}$	$6.9 \cdot 10^{-6}$	$1.6 \cdot 10^{-5}$	$4.8 \cdot 10^{-5}$	$3.7 \cdot 10^{-4}$
M_{v_2}	$1.5 \cdot 10^{-6}$	$3.3 \cdot 10^{-6}$	$6.9 \cdot 10^{-6}$	$1.6 \cdot 10^{-5}$	$4.6 \cdot 10^{-5}$	$3.4 \cdot 10^{-4}$

Note the very small values of the three measures that apply also for large values of the angle ω; that proves that second order terms are not required in typical imaging situations.

When is first order exact?

If $\mu = 0$ in (4.6) (for instance when $\omega = 0^o$), then the second order correction is zero and all the three above measures are null. With $\mu = 0$ the homography reduces to an affine transformation, i.e. it is linear. This illustrates the general result that if the homography is affine the first order analysis is exact.

Generally, the $\mathtt{H}$ matrix can be decomposed into the product of matrices representing *linear* (affine) and *non-linear* (projective) transformations on in-homogeneous coordinates as:

$$\mathtt{H} = \mathtt{AP}$$

where:

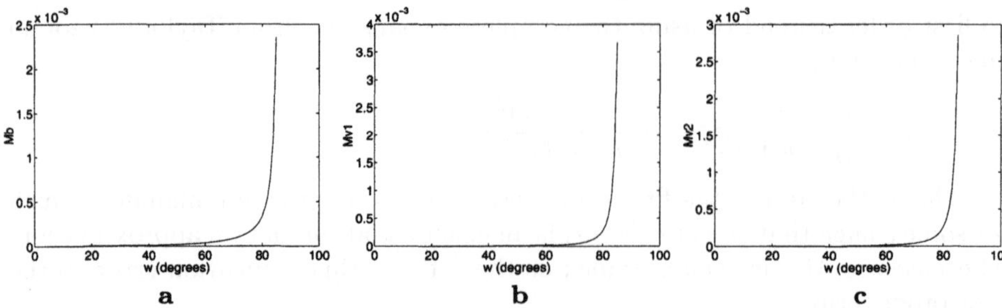

Fig. 4.3. Comparison of measures of fitness of first order approximation vs. second order: **a** bias measure M_b; **b** first variance measure M_{v_1}; **c** second variance measure M_{v_2}.

$$\mathtt{A} = \begin{pmatrix} \alpha_{11} & \alpha_{12} & h_{13} \\ \alpha_{21} & \alpha_{22} & h_{23} \\ 0 & 0 & 1 \end{pmatrix}, \quad \mathtt{P} = \begin{pmatrix} 1 & 0 & 0 \\ 0 & 1 & 0 \\ h_{31} & h_{32} & 1 \end{pmatrix}$$

with $\alpha_{ij} = h_{ij} - h_{3j}h_{i3}$ for $i, j = 1, 2$.

If $\mathtt{H}$ is purely an affinity (linear on inhomogeneous coordinates) then $\mathtt{P}$ is the identity and the first order theory exact.

In the previous section, the first order error propagation theory was proved to be good enough for typical values in the case of the mono-dimensional camera model. In the following, the first order analysis is extended to the bi-dimensional camera model in the case of uncertainty affecting the projection process also. Then, in section 4.2.3 its correctness is proved via Monte Carlo statistical tests.

4.2.2 Computing uncertainties

There may be errors in the world and image points used to compute the homography, and there may be errors in the image points back-projected to make world measurements. All of these sources of uncertainty must be taken into account in order to compute a cumulative uncertainty for the world measurement (cf. Fig. 1.2).

In this section, formulae to compute the uncertainty for planar measurements under various error situations are developed. The first order analysis is assumed sufficient.

All the computation image- and world-points are assumed to be measured with error modelled as a bi-dimensional Gaussian noise process (see Fig. 3.13). $\Lambda_{\mathbf{x}_i}$ and $\Lambda_{\mathbf{X}_i}$ are the covariance matrices of the image computation point $\mathbf{x}_i$ and the world computation point $\mathbf{X}_i$ respectively. The two sources of error (the uncertainty on the homography and the uncertainty in image point localization) are first considered to operate separately, and finally they are merged in order to compute an unique formula embracing both cases.

Uncertainty of the homography H given uncertain computation points.

The covariance of the homography H estimated from n image-world point correspondences is computed here.

From section 4.1.1 we seek the eigenvector $\mathbf{h}$ with smallest eigenvalue λ of the 3×3 matrix $\mathtt{A}^\top \mathtt{A}$. But if the n computation correspondences are affected by a localization error also, the A matrix will be uncertain and hence the 9-vector $\mathbf{h}$. This vector is then characterized by a 9×9 covariance matrix $\Lambda_{\mathbf{h}}$ computed as shown in Appendix A.

The algorithm in Appendix A has a double advantage over other methods such as [19, 38] which require the inverse of $\mathtt{A}^\top \mathtt{A}$ in order to compute $\Lambda_{\mathbf{h}}$. These methods are poorly conditioned if only four correspondences are used, or if $n > 4$ and the correspondences are (almost) noise-free. In both cases the $\mathtt{A}^\top \mathtt{A}$ matrix is singular and thus not invertible. Since the derivation of (A.2) does not involve the inverse, it is well conditioned in both cases.

Uncertainty in point localization.
This section lists the formulae used to compute the uncertainty for measurements under various error situations. The first order analysis is assumed sufficient. The uncertainty in the homography is computed as described in the previous section.

Notation.
New notation is introduced here to simplify the formulae. Equation (3.3) can be rewritten as:

$$\mathbf{X} = \mathtt{B}\mathbf{h}$$

where B is a 3×9 matrix in the form:

$$\mathtt{B} = \begin{pmatrix} \mathbf{x}^\top & \mathbf{0}^\top & \mathbf{0}^\top \\ \mathbf{0}^\top & \mathbf{x}^\top & \mathbf{0}^\top \\ \mathbf{0}^\top & \mathbf{0}^\top & \mathbf{x}^\top \end{pmatrix} \tag{4.10}$$

In the following we will determine the 3×3 covariance matrix $\Lambda_{\mathbf{X}}$ of the homogeneous world point $\mathbf{X}$. The conversion to a 2×2 covariance matrix $\Lambda_{\mathbf{X}}^{2\times2}$ for inhomogeneous coordinates is given by: $\Lambda_{\mathbf{X}}^{2\times2} = \nabla f \Lambda_{\mathbf{X}} \nabla f^\top$ where $\mathbf{X} = (X, Y, W)^\top$ and:

$$\nabla f = 1/W^2 \begin{pmatrix} W & 0 & -X \\ 0 & W & -Y \end{pmatrix}$$

f is the function mapping homogeneous to inhomogeneous coordinates. The opposite conversion, from inhomogeneous to homogeneous coordinates, for an image point $\mathbf{x}$ is simply given by:

$$\Lambda_{\mathbf{x}} = \begin{pmatrix} \Lambda_{\mathbf{x}}^{2\times2} & \mathbf{0} \\ \mathbf{0}^\top & 0 \end{pmatrix}$$

It is easy to prove that, given two homogeneous 3-vectors $\mathbf{x}_1$ and $\mathbf{x}_2$ related by:

$$\mathbf{x}_2 = \boldsymbol{\lambda}\mathbf{x}_1 \tag{4.11}$$

with $\boldsymbol{\lambda}$ a scale factor, and their associated 3×3 covariance matrices $\Lambda_{\mathbf{x}_1}$ and $\Lambda_{\mathbf{x}_2}$, then $\Lambda_{\mathbf{x}_2}^{2\times2} = \Lambda_{\mathbf{x}_1}^{2\times2}$ where $\Lambda_{\mathbf{x}_1}^{2\times2}$ and $\Lambda_{x_2}^{2\times2}$ are the corresponding inhomogeneous covariance matrices. In fact from (4.11):

$$\Lambda_{\mathbf{x}_2} = \lambda^2 \Lambda_{\mathbf{x}_1} \quad \text{and} \quad \nabla f_2 = \frac{1}{\lambda}\nabla f_1$$

Since from first order error analysis $\Lambda_{\mathbf{x}_2}^{2\times2} = \nabla f_2 \Lambda_{\mathbf{x}_2} \nabla f_2^{\top}$, by substitution:

$$\Lambda_{\mathbf{x}_2}^{2\times2} = \frac{1}{\lambda}\nabla f_1 \lambda^2 \Lambda_{\mathbf{x}_1} \frac{1}{\lambda}\nabla f_1{}^{\top}$$

and since $\nabla f_1 \Lambda_{\mathbf{x}_1} \nabla f_1{}^{\top} = \Lambda_{\mathbf{x}_1}^{2\times2}$ then $\Lambda_{\mathbf{x}_2}^{2\times2} = \Lambda_{\mathbf{x}_1}^{2\times2}$. Therefore, as expected, the inhomogeneous covariance matrices are identical and one of the homogeneous covariances is the scaled version (with positive scale factor) of the other.

Uncertainty in $\mathbf{X}$*, given an uncertain* $\mathtt{H}$ *and exact* $\mathbf{x}$.
If noise is assumed only on the homography and the input point $\mathbf{x}$ is exact, then the covariance of the corresponding world point $\mathbf{X}$ is:

$$\Lambda_{\mathbf{X}} = \mathtt{B}\Lambda_{\mathbf{h}}\mathtt{B}^{\top} \tag{4.12}$$

Uncertainty in $\mathbf{X}$*, given an exact* $\mathtt{H}$ *and uncertain* $\mathbf{x}$.
Instead, if noise is assumed only in the input image point and the homography is exact then:

$$\Lambda_{\mathbf{X}} = \mathtt{H}\Lambda_{\mathbf{x}}\mathtt{H}^{\top} \tag{4.13}$$

Uncertainty in $\mathbf{X}$*, given an uncertain* $\mathtt{H}$ *and uncertain* $\mathbf{x}$.
If noise is assumed in both the homography $\mathtt{H}$ and the input point $\mathbf{x}$ then the covariance $\Lambda_{\mathbf{X}}$ is given by the sum of the previous two equations (4.12, 4.13):

$$\Lambda_{\mathbf{X}} = \begin{pmatrix} \mathtt{B} \,\vdots\, \mathtt{H} \end{pmatrix} \begin{pmatrix} \Lambda_{\mathbf{h}} & \vdots & 0 \\ \cdots & \cdots & \cdots \\ 0 & \vdots & \Lambda_{\mathbf{x}} \end{pmatrix} \begin{pmatrix} \mathtt{B}^{\top} \\ \cdots \\ \mathtt{H}^{\top} \end{pmatrix} \tag{4.14}$$

if $\mathtt{H}$ and $\mathbf{x}$ are statistically independent.

Uncertainty in distance measurement.
The distance d between two world points $\mathbf{U}_1$ and $\mathbf{U}_2$ (inhomogeneous) also has an associated uncertainty.

Let the two end points of the image of the segment to be measured be $\mathbf{u}_1 = (u_1, v_1)^{\top}$ and $\mathbf{u}_2 = (u_2, v_2)^{\top}$ (in inhomogeneous coordinates). The two corresponding points on the world plane in homogeneus coordinates are: $\mathbf{X}_1 = (X_1, Y_1, W_1)^{\top}$ and $\mathbf{X}_2 = (X_2, Y_2, W_2)^{\top}$ where $\mathbf{X}_i = \mathtt{H}(u_i, v_i, 1)^{\top}$. In inhomogeneus coordinates they are: $\mathbf{U}_i = (U_i, V_i)^{\top}$ with $U_i = X_i/W_i$ and $V_i = Y_i/W_i$. Then the Euclidean distance between the two points $\mathbf{U}_1$ and $\mathbf{U}_2$ on the world plane is:

$$d = \sqrt{(U_1 - U_2)^2 + (V_1 - V_2)^2}$$

and its uncertainty is computed below.

Uncertainty on point distance, given an uncertain $\mathtt{H}$ *and exact* $\mathbf{u}_1, \mathbf{u}_2$.
It can be shown that the gradient of the distance d with respect to $\mathbf{h}$ is:

$$\frac{\partial d}{\partial \mathbf{h}} = \frac{1}{d}(\mathbf{U}_1 - \mathbf{U}_2)^\top (\tilde{\mathtt{A}}_1 - \tilde{\mathtt{A}}_2)$$

where the 2×9 matrices $\tilde{\mathtt{A}}_1$ and $\tilde{\mathtt{A}}_2$ are:

$$\tilde{\mathtt{A}}_i = \frac{1}{W_i}\begin{pmatrix} u_i & v_i & 1 & 0 & 0 & 0 & -u_i U_i & -v_i U_i & -U_i \\ 0 & 0 & 0 & u_i & v_i & 1 & -u_i V_i & -v_i V_i & -V_i \end{pmatrix}$$

If noise is assumed only in the homography, then for the first order analysis, the variance of d is given by:

$$\sigma_d^2 = \frac{\partial d}{\partial \mathbf{h}} \Lambda_{\mathbf{h}} \frac{\partial d}{\partial \mathbf{h}}^\top \tag{4.15}$$

Uncertainty in distance, given exact $\mathtt{H}$ *and uncertain* $\mathbf{u}_1$ *and* $\mathbf{u}_2$.
The gradient of the distance with respect to the 4-vector $\boldsymbol{\zeta} = (u_1, v_1, u_2, v_2)^\top$ is given by:

$$\frac{\partial d}{\partial \zeta} = \frac{1}{d}(\mathbf{U}_1 - \mathbf{U}_2)^\top \left(\mathtt{B}_1 \vdots \mathtt{B}_2\right)$$

where the matrices $\mathtt{B}_1$ and $\mathtt{B}_2$ are defined as follows:

$$\mathtt{B}_i = \frac{1}{W_i}\begin{pmatrix} h_{11} - h_{31}U_i & h_{12} - h_{32}U_i \\ h_{21} - h_{31}V_i & h_{22} - h_{32}V_i \end{pmatrix}$$

If noise is assumed only in the input points $\mathbf{u}_1, \mathbf{u}_2$, and the two points are independent, then the variance σ_d^2 is:

$$\sigma_d^2 = \frac{\partial d}{\partial \zeta} \begin{pmatrix} \Lambda_{\mathbf{u}_1} & \vdots & 0 \\ \cdots & \cdots & \cdots \\ 0 & \vdots & \Lambda_{\mathbf{u}_2} \end{pmatrix} \frac{\partial d}{\partial \zeta}^\top \tag{4.16}$$

Uncertainty in distance, given uncertain $\mathtt{H}$ *and uncertain* $\mathbf{u}_1$ *and* $\mathbf{u}_2$.
Finally, if both the homography $\mathtt{H}$ and the input points $\mathbf{u}_1, \mathbf{u}_2$ are uncertain then the variance σ_d^2 is given by the sum of (4.15) and (4.16):

$$\sigma_d^2 = \left(\frac{\partial d}{\partial \mathbf{h}} \vdots \frac{\partial d}{\partial \zeta}\right) \begin{pmatrix} \Lambda_{\mathbf{h}} & \vdots & 0 & \vdots & 0 \\ \cdots & \cdots & \cdots & \cdots & \cdots \\ 0 & \vdots & \Lambda_{\mathbf{u}_1} & \vdots & 0 \\ \cdots & \cdots & \cdots & \cdots & \cdots \\ 0 & \vdots & 0 & \vdots & \Lambda_{\mathbf{u}_2} \end{pmatrix} \begin{pmatrix} \frac{\partial d}{\partial \mathbf{h}}^\top \\ \cdots \\ \frac{\partial d}{\partial \zeta}^\top \end{pmatrix} \tag{4.17}$$

with $\Lambda_{\mathbf{h}}$ the 9×9 covariance of the homography and $\Lambda_{\mathbf{u}_1}$ and $\Lambda_{\mathbf{u}_2}$ the 2×2 inhomogeneous covariance matrices of the input image points.

4.2.3 Validation of uncertainty analysis

The previous sections have described a complete theory for computing:

- the ellipse of uncertainty for the localization of a world plane point;
- the uncertainty range of the distance between two world points given uncertain homography computation points and uncertain image input points.

This section demonstrates that, for typical imaging situations, this linear approximation is sufficiently accurate. That has already been proven analytically in the mono-dimensional case. Now the Monte Carlo statistical test is applied to the bi-dimensional camera model theory. In all of the following, error is assumed to be in both computation points and measurement points.

Test on uncertainty of point localization.
The performed *Monte Carlo* test is described in Table 4.1.

• Repeat N times – Generate $n \geq 4$ image computation points (Gaussian distributed); – Generate $n \geq 4$ world computation points (Gaussian distributed); – Compute the homography relating the two sets of points. – Generate an image point $\mathbf{x}$ (Gaussian distributed); – Back-project the point $\mathbf{x}$ onto the world plane point $\mathbf{X}$. • Compute the statistical uncertainty ellipse of the distribution of points $\mathbf{X}$ and compare it with the analytical uncertainty.

Table 4.1. Monte Carlo test.

Figure 4.4**a** shows a comparison between the covariance ellipse obtained by the first order analysis and that obtained by a Monte Carlo evaluation of the actual non-linear homography mapping. The number N of iterations involved in each simulation is 10 000. Note that the predicted ellipse and the simulated one are almost overlapping. These figures are obtained using parameters related to a real situation.

Similar test has been performed in section 3.4 for a homographic mapping between two generic planes, using synthetic data.

Increasing the number of computation points.
Figure 4.4b shows what happens if the number of computation points used to estimate the homography matrix increases. Six ellipses are drawn, three have been predicted by the analytical theory and another three obtained from the Monte Carlo test. For each of these three cases it is very hard to distinguish between the simulated and the predicted ellipses.

Furthermore, as expected, increasing the number of computation points increases the accuracy of the $\mathtt{H}$ matrix and also the accuracy of the final position of points $\mathbf{X}$ (smaller ellipses).

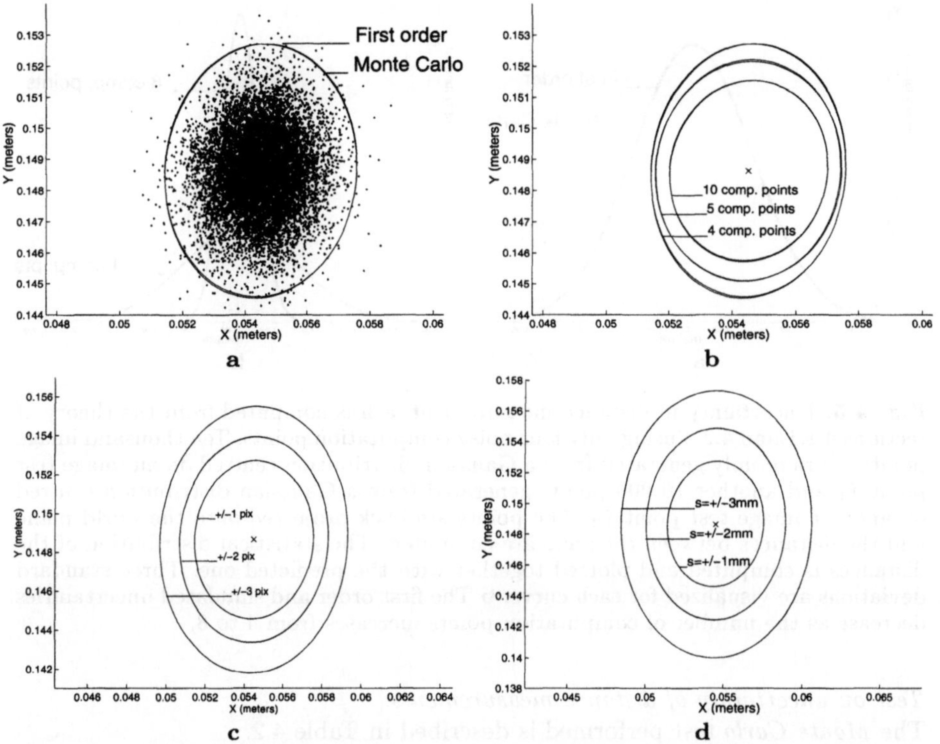

Fig. 4.4. Uncertainty in point localization. **a** $\mathtt{H}$ is computed from the theory of sections 4.1.1 and 4.2.2 using only four computation points. Ten thousand image points are randomly generated from a Gaussian distribution centred on an image test point $\overline{x}$ and then back-projected onto the world plane. The statistical covariance ellipse of the world points is computed and plotted, together with the predicted one. Three standard deviations are visualized for each uncertainty ellipse. **b** The areas of the first order and simulated uncertainty ellipses decrease as the number of computation points increase from four to five to ten as expected from the theory. **c** Areas of the predicted ellipses and of the simulated ones increase as the uncertainty of the image computation points increases. **d** Areas of the predicted ellipses and of the simulated ones increase as the uncertainty of the world computation points increases. Three standard deviations (99.7% of probability level) are visualized in all diagrams.

Increasing the standard deviation of the image computation points.
Figure 4.4**c** shows how the area of the uncertainty ellipses increases with the uncertainty of the image computation points. Three predicted ellipses are drawn, together with the corresponding simulated ones. The standard deviation of the noise in the world points varies from 1 pixel to 3 pixels.

Increasing the standard deviation of the world computation points.
The increase of the uncertainty ellipses with uncertainty in the localization of the world computation points is shown in Fig. 4.4**d**. The three predicted ellipses and the corresponding simulated ones are also drawn in this case. The standard deviation of the noise in the world points varies from 1 mm to 3 mm.

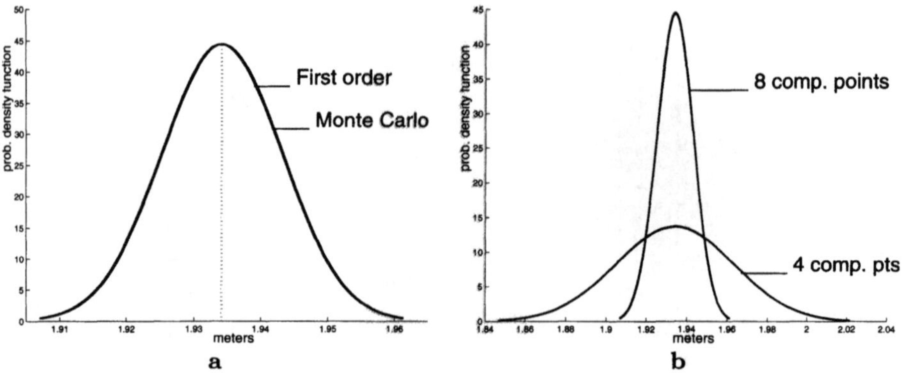

Fig. 4.5. Uncertainty in distance measurement. **a** H is computed from the theory of sections 4.1.1 and 4.2.2 using only four noisy computation points. Ten thousand image points are randomly generated from a Gaussian distribution centred on an image test point $\overline{x}_1$ and another 10 000 points generated from a Gaussian distribution centred on another image test point $\overline{x}_2$. The points are back-projected onto the world plane and the distances between the pairs are computed. The statistical distribution of the distances is computed and plotted together with the predicted one. Three standard deviations are visualized for each curve. **b** The first order and simulated uncertainties decrease as the number of computation points increases from 4 to 8.

Test on uncertainty of distance measurements.
The *Monte Carlo* test performed is described in Table 4.2.

Figure 4.5 demonstrates the uncertainty in distance measurements. A comparison between the Gaussian distribution obtained by the first order analysis and, again, the distribution obtained by a Monte Carlo evaluation of the non-linear homography mapping is shown. The predicted curve and the simulated one are again indistinguishable.

- Repeat N times
  - Generate $n \geq 4$ image computation points (Gaussian distributed);
  - Generate $n \geq 4$ world computation points (Gaussian distributed);
  - Compute the homography relating the two sets of points.
  - Generate an image point $\mathbf{x}_1$ (Gaussian distributed);
  - Back-project the point $\mathbf{x}_1$ onto the world plane point $\mathbf{X}_1$.
  - Generate an image point $\mathbf{x}_2$ (Gaussian distributed);
  - Back-project the point $\mathbf{x}_2$ onto the world plane point $\mathbf{X}_2$.
  - Compute the distance d between $\mathbf{X}_1$ and $\mathbf{X}_2$.
- Compute the statistical variance of the distribution of distances d and compare it with the predicted variance.

Table 4.2. Monte Carlo test.

Increasing the number of computation points.
Figure 4.5**b** illustrates what happens if the number of computation points used to estimate the homography matrix increases.

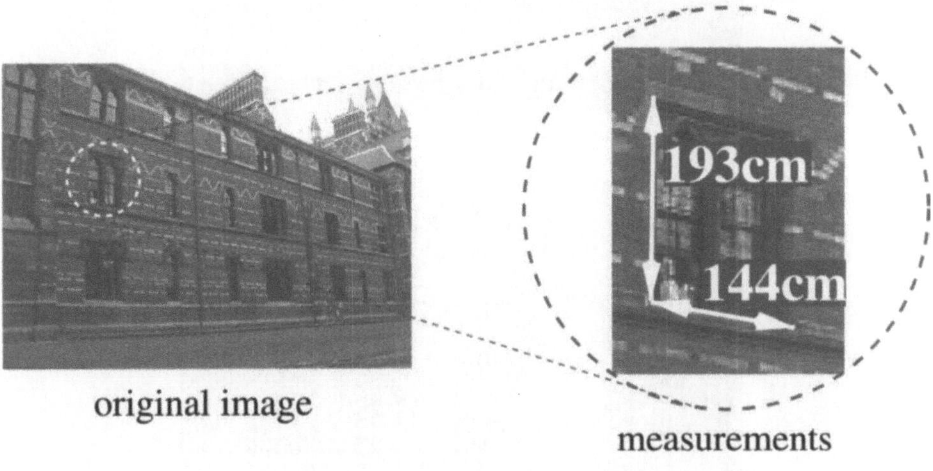

Fig. 4.6. A plane measuring device: measuring world distances from images. The real dimensions of the window are measured directly on the image.

Four curves are drawn, two have been predicted by the analytical theory and the other two have been obtained from the Monte Carlo test. As expected, increasing the number of computation points increases the accuracy of the distance d (smaller output standard deviation, sharper peaks).

Figures 4.4 and 4.5 are obtained using data from images of real scenes. In all the previous examples, the simulated ellipses almost exactly overlap the analytically predicted ones, which justifies the use of the first order theory.

4.3 Application – a plane measuring device

In this section a typical application of the described plane metrology algorithm and the uncertainty theory is presented.

4.3.1 Description

The device shown here is meant to take distance measurements on real-world planes, such as floors, walls and doors, just using images acquired with conventional video cameras (see Fig. 4.6). The entire measurement process can be split into two main stages: the calibration stage and the measuring stage.

Calibration stage.
First of all, computation of the transformation between the viewed plane and the image plane is necessary. This is achieved via the knowledge of a set of world-image point correspondences as already discussed.

Once we have defined how the n chosen world-plane computation points are projected onto the image it is possible to apply the theory of sections 4.1 and 4.2 to compute the H matrix and its covariance. The choice of computation points

Fig. 4.7. A Plane Measuring Device: Graphical User Interface, example of the measurement stage. It involves an easy-to-use interface; the user is only required to select the two end points of the segment to measure and the distance measure and related uncertainty pop up.

is not completely arbitrary; in fact their number and location with respect to the camera have a significant effect on measurement uncertainties.

In section 4.3.2 some examples are provided to show the variation of the uncertainties according to a variation in the number or location of the computation points.

Measurement stage.

Once the homography has been computed two points on the image are selected; they are back-projected onto the world plane and the distance between them computed and shown (see Fig. 4.7). The uncertainty in the distance between them is also computed using the formulae presented.

Once the homography is known it is possible to retrieve other useful information such as parallelism of lines lying on a plane. This application is as follows: draw a line, draw the parallel line (parallel in the world and not in the image) passing through a chosen point, and compute the orthogonal distance between the two lines.

The strength of this device is in its ease of use; the operator interacts with the system just via a friendly graphic interface, both during the calibration stage and the measuring one. The measurement queries are sent to the system via a pointing device such as a mouse.

4.3.2 Examples

In this section a number of examples are presented showing possible uses of the described theory, and demonstrating the applicability of the uncertainty analysis to real images.

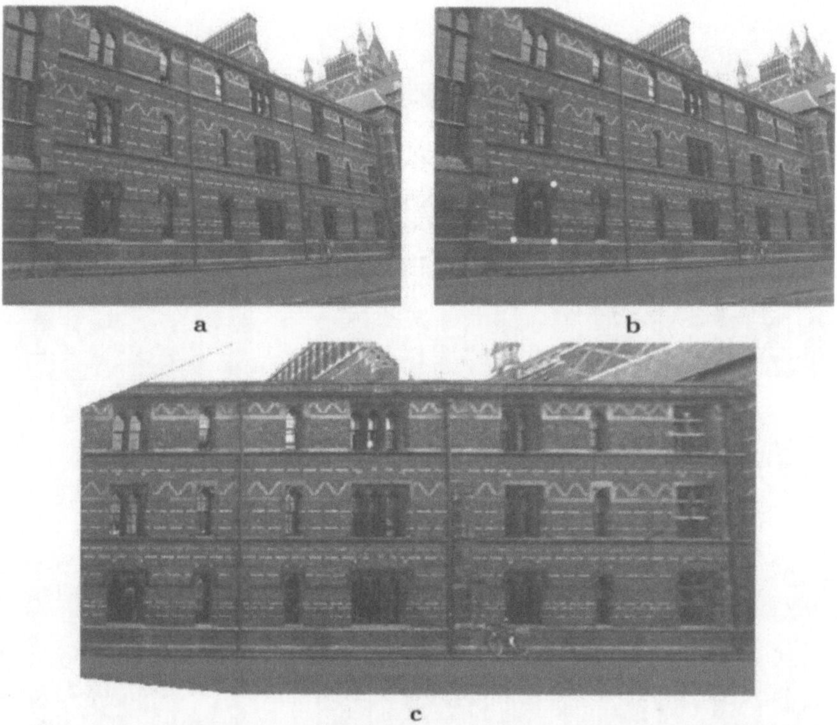

Fig. 4.8. Rectification of a planar surface: **a** original image, Keble College, Oxford; **b** the relative location of the four corners of the marked window have been used as reference to compute the image-to-world plane homography; **c** rectified image; parallel lines in the world are parallel in the rectified image and angles are preserved.

It is shown that the ground truth measurements always lie within the estimated error bounds. Furthermore, the utility of the analysis is illustrated. The covariance expression *predicts* uncertainty given the number of image-world computation points and their distribution. It is thus possible to decide where further reference correspondences are required in order to achieve a particular desired measurement accuracy.

Creating new views of planar surfaces.

Rectification of a planar surface.

Once the homography between a world plane and the image has been estimated, then it is easy to perform a *fronto-parallel* transformation of the image. From a geometric point of view, this operation means the synthesis of a new image, the one seen from a new camera with its acquisition plane parallel to the world plane.

Figure 4.8 shows an image of a college wall which is warped onto a fronto-parallel view. Note that in the warped image (Fig. 4.8c) perspective distortion is corrected: angles and ratios of distances are recovered correctly.

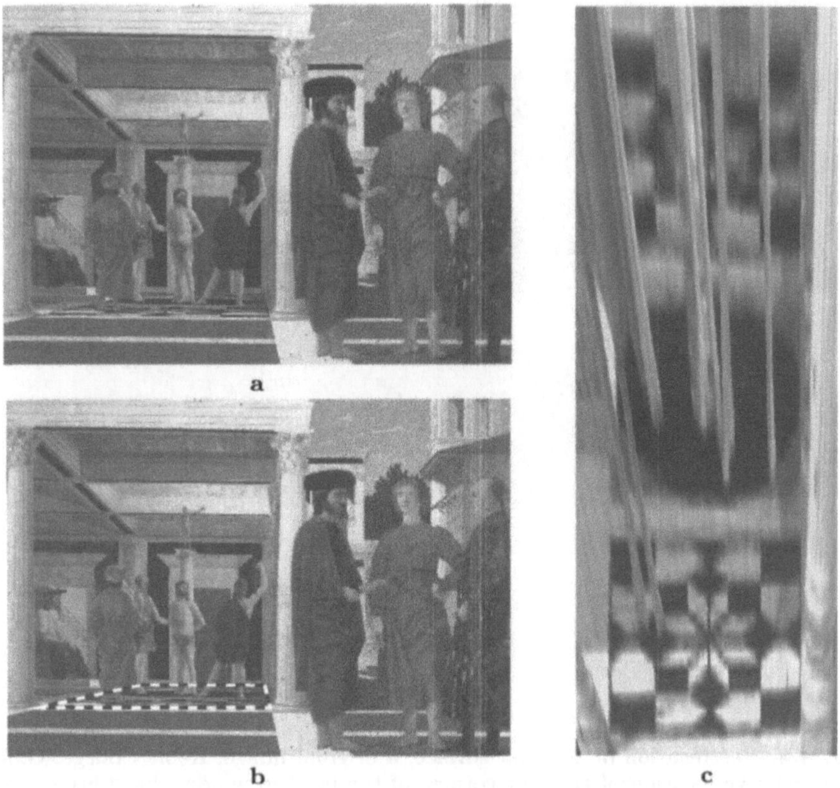

Fig. 4.9. Rectification of a plane in a painting. **a** The painting *Flagellazione* (approx. 1453), by Piero della Francesca (1416–1492), Galleria Nazionale delle Marche, Urbino, Italia. Courtesy of Ministero per i Beni e le Attività Culturali. **b** Original image with part of the floor highlighted. **c** Rectified image of the highlighted area. Notice that the beautiful geometrical tile pattern is repeated twice; the one behind the figure of Christ is barely visible in the original painting.

Rectification of a plane in a painting.
Figure 4.9 shows another example of plane rectification. Figure 4.9**a** is a painting by the Italian Renaissance painter Piero della Francesca (1416–1492). The rectification is possible as a result of strict adherence to Renaissance perspective rules by the artist. The painting represents the geometry of the scene almost exactly as it would be captured by a perspective camera, thus the warping technique described is valid.

Figure 4.9**c** shows the rectified view of the area of the floor highlighted in Fig. 4.9**b**, where Christ is standing. Note that the geometric pattern, barely visible in the original image, is clear at the bottom of the rectified one [67]. A second, identical pattern is visible (despite occlusions) on the top of the rectified image.

The image-to-world homography has been computed from the assumption of a square floor pattern.

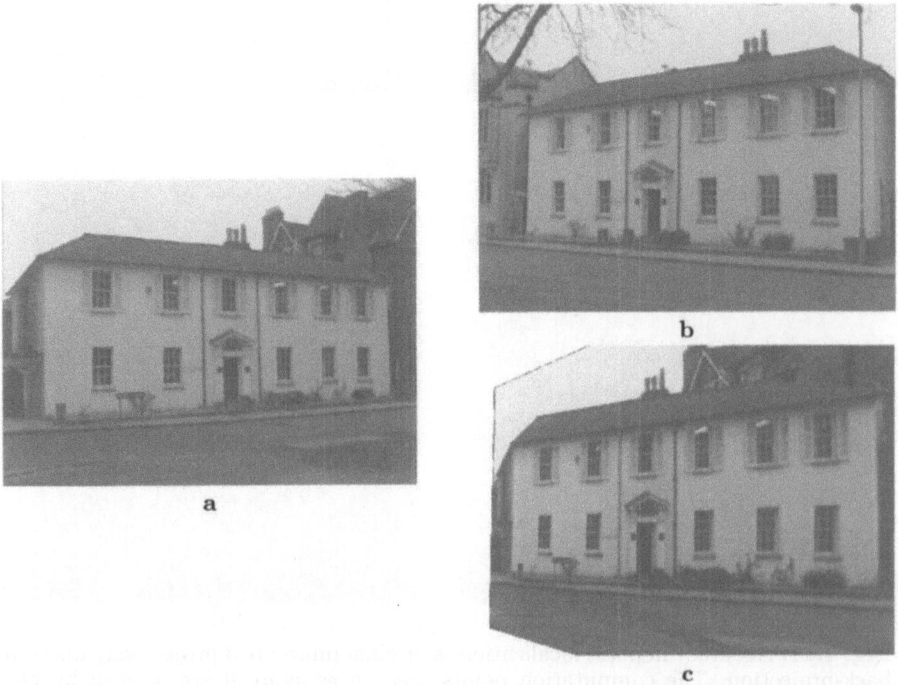

Fig. 4.10. Warping planes between images. a,b Original images from two different points of view; Crystallography Labs, Oxford. c First image warped onto the second by the computed inter-image homography.

Warping planes between images.
If the inter-image homography relating two images is known, it is possible to warp one image onto the other.

An example is shown in Fig. 4.10 where two pictures of the same wall have been taken from two different viewpoints (the camera has undergone a translational and rotational movement). The inter-image homography has been estimated from corresponding feature points on the wall and the first image warped onto the second one. All the points which do not lie on the facade are mapped into unexpected positions (Fig. 4.10**c**). The parallax effect shown by such points can be used for three-dimensional structural computation (see Chapter 6).

Accuracy in point localization.

Varying the location of the computation points.
Figure 4.11 shows an example of an indoor scene. Fig. 4.11**a** is the original image and in Figs. 4.11**b–d** an *uncertainty ellipse map* is superimposed to show how the uncertainty ellipses change their shape according to a change only in the location of the computation points. The H matrix is computed using four computation points in all cases. The covariances used for the computations are:

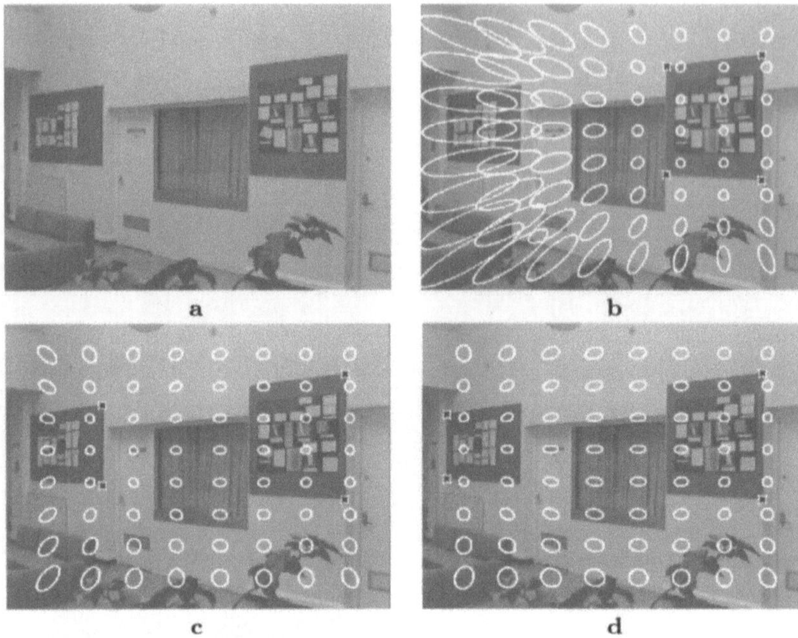

Fig. 4.11. Accuracy in point localization: **a** original image; **b-d** projectively unskewed back-projection. The computation points used to estimate H are marked by black asterisks. The location of the four computation points is different in each image. The uncertainty ellipses shown for test points are at the nine standard deviation level.

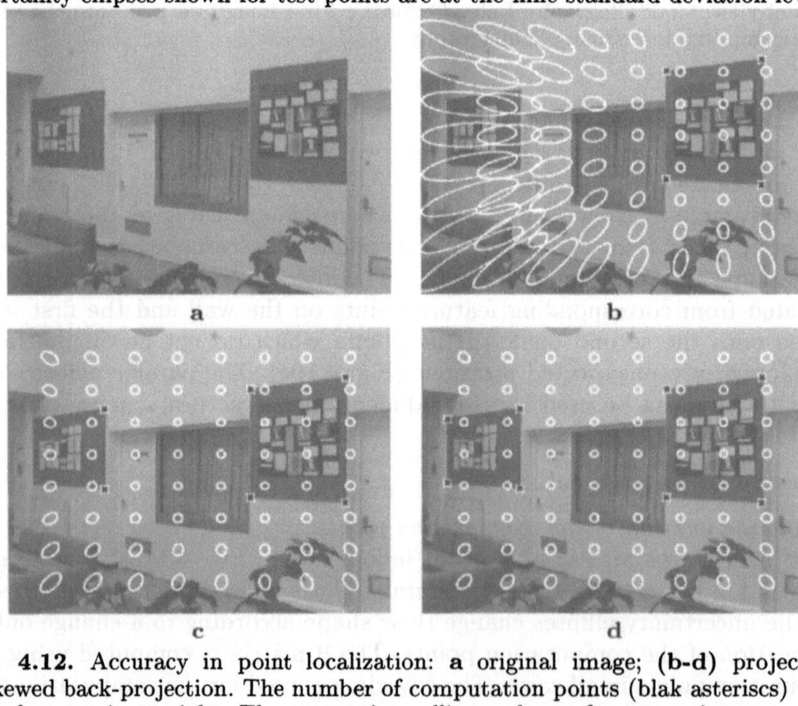

Fig. 4.12. Accuracy in point localization: **a** original image; **(b-d)** projectively unskewed back-projection. The number of computation points (blak asteriscs) varies from four to six to eight. The uncertainty ellipses shown for test points are at the nine standard deviation level for clarity.

$$\Lambda_x = \begin{bmatrix} 1 & 0 \\ 0 & 1 \end{bmatrix}$$

(in pixel^2) in the image and:

$$\Lambda_{\mathbf{X}} = \begin{bmatrix} 0.25 & 0 \\ 0 & 0.25 \end{bmatrix}$$

(in cm^2) in the world.

Several test points are shown with their uncertainty ellipses. Note that as the distance of the test point to computation points increases, the uncertainty increases. More spatially homogeneous uncertainties are achieved by distributing the computation points across the scene.

Varying the number of computation points.
Figure 4.12 shows the same scene as before. Figure 4.12**a** is the original image and Figs 4.12**b-d** show the uncertainty ellipses map.

This time we modify the number of computation points. The `H` matrix is computed using four points in Fig. 4.12**b**, six point in Fig. 4.12**c** and eight points in Fig. 4.12**d**. The covariances used for the computations are as before:

$$\Lambda_x = \begin{bmatrix} 1 & 0 \\ 0 & 1 \end{bmatrix} \quad \Lambda_{\mathbf{X}} = \begin{bmatrix} 0.25 & 0 \\ 0 & 0.25 \end{bmatrix}$$

The same test points are shown with their uncertainty ellipses. Smaller uncertainties are obtained by increasing the number of computation points.

Accuracy of distances.
A similar analysis is now conducted using images of an outside wall, computing distances instead of point locations.

Varying the number of computation points.
Figure 4.13**b–d** show length measurements for a homography computed from four, six and eight correspondences. The covariances used for the computations are:

$$\Lambda_x = \begin{bmatrix} 1 & 0 \\ 0 & 1 \end{bmatrix} \quad \Lambda_{\mathbf{X}} = \begin{bmatrix} 1 & 0 \\ 0 & 1 \end{bmatrix}$$

Measurements farther from the set of computation points present a larger uncertainty. Again, increasing the number of computation points reduces the uncertainties in all the measurements. Note that all the ranges of estimated measurements include the actual window width (ground truth is 139 cm).

Different views, same computation points.
Figure 4.14 illustrates that the uncertainties also depend on the observer viewpoint. The computation points for both images are in the same zone of the image frame but the second image is affected by a more severe perspective distortion. In both cases the ground truth lies within the predicted measurement range, but this is larger in the second view, as expected.

The figure also illustrates the computation of parallel world lines (see Table 4.3). Once a line is selected in the image, the one-parameter family of lines parallel to it on the world plane is computed from the estimated `H`. When one of them is fixed, the orthogonal distance between them is computed and shown.

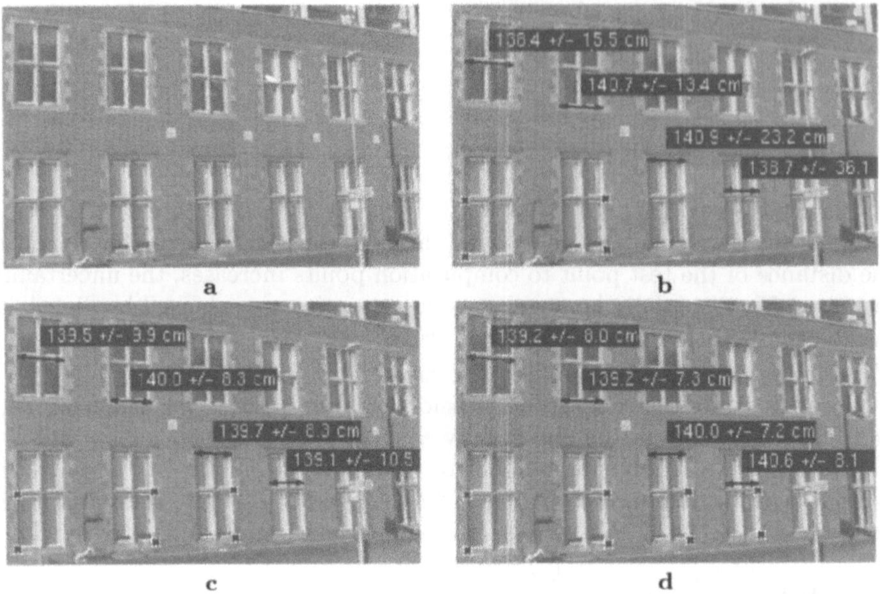

Fig. 4.13. Accuracy of distances: **a original image; b-d** length measurements based on a homography computed from the points marked by black asterisks. The uncertainty bound is ±3 standard deviations. The actual width is 139 cm.

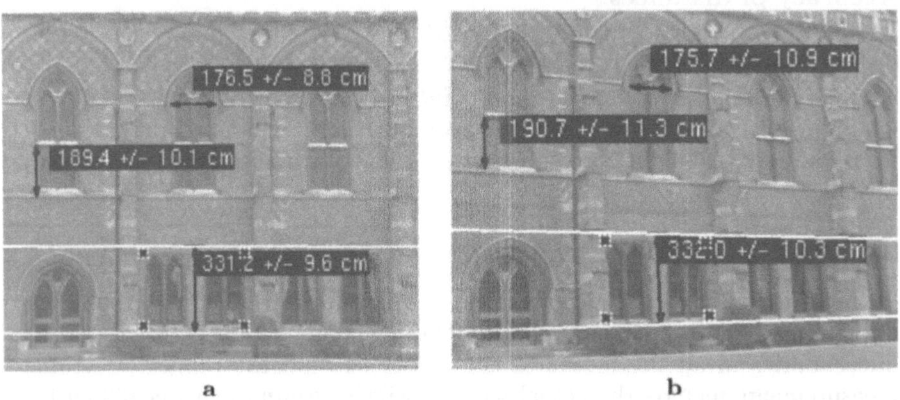

Fig. 4.14. Different views, same computation points: **a** and **b** two images of Keble College, Oxford. The computation points are the same, but the viewpoint distortion is more severe in **b**. This is reflected in the larger (3 standard deviation) uncertainties. The actual width of the upper windows is 176 cm. Note the computed parallel lines.

Different views, different computation points.
Figure 4.15 shows, again, two different views of a wall. Four different computation points are used in the two images. All the measurements are taken between parallel lines and although the angle between camera and wall plane is large, the parallel lines are correctly computed.

A line $\mathbf{l}$ on the image defines a back-projected plane π_l in space. The intersection of the plane π_l with a distinguished world planar surface π defines a line $\mathbf{L}$.
Given the homography H betweeen the image plane and the plane π and the line $\mathbf{l}$ on the image, the one-parameter family of lines parallel to $\mathbf{L}$ in the world is defined.
In fact, given a line $\mathbf{l}$ and a point $\mathbf{x}$ in the image, we can find the line $\mathbf{l}'$ through $\mathbf{x}$ whose back-projection $\mathbf{L}'$ is parallel to $\mathbf{L}$ on the world plane. This is simply $\mathbf{l}' = \mathbf{x} \times \mathbf{v}$ where $\mathbf{v}$ is the vanishing point for that direction; $\mathbf{v}$ is given by $\mathbf{v} = \mathtt{H}^{-1}\mathtt{R}\mathtt{H}^{-\top}\mathbf{l}$ with R the following filter matrix:

$$\mathtt{R} = \begin{pmatrix} 0 & 1 & 0 \\ -1 & 0 & 0 \\ 0 & 0 & 0 \end{pmatrix}$$

Once a pair of lines $\mathbf{l}, \mathbf{l}'$ is selected, the orthogonal world distance between them can be easily computed by applying (3.3).

Table 4.3. Computing images of parallel lines and measuring their distance.

Note that the distance measurements are *invariant* to the choice of computation points, and the ground truth is always in the uncertainty range returned by the system (see caption).

Fig. 4.15. Different views, different computation points. **a** and **b** two images of the Crystallography Labs, Oxford. The actual height of all the windows is 174 cm, the door width is 100*cm*. Notice the correct computation of parallel lines.

Mosaicing and measuring.
If the inter-image homography between two views is known, then one image can be warped and stitched to the other, thus making a mosaic image. Note, though, that in general only the points lying on the same plane are registered correctly by the homography, while the ones off the plane are warped into unexpected positions [11] because of the parallax effect.

In the case of a purely rotating camera, no parallax can be detected and therefore all the points are registered correctly by the homography. An example of a mosaic created from a rotating camera is shown in Fig. 4.16. Figure 4.16**a** and **b** are two different images of an indoor scene taken from the same point of view (no parallax effect). The inter-image homography is computed from the overlapping area between the two images and the second image warped

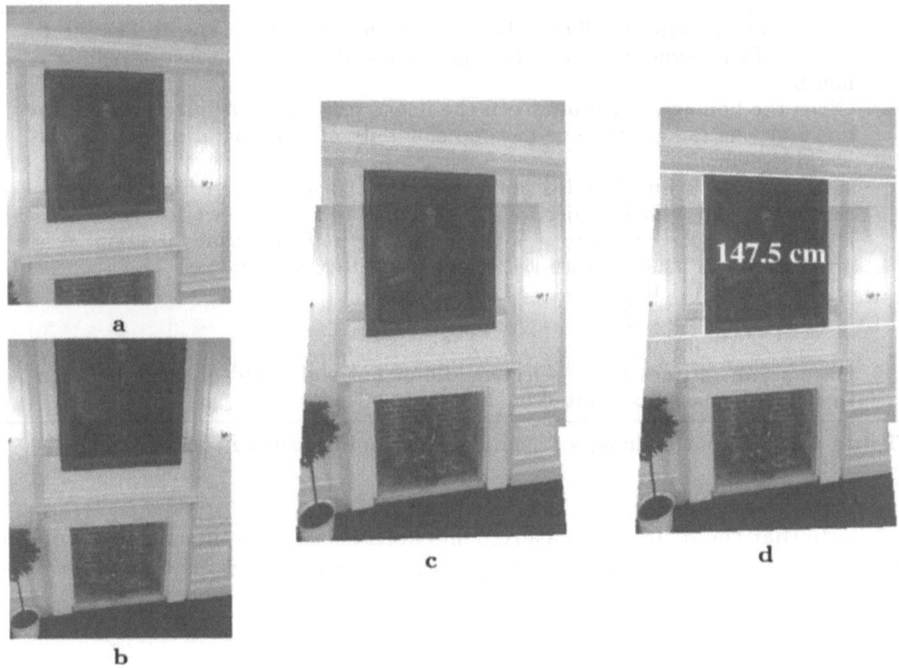

Fig. 4.16. Mosaicing two images: **a** and **b** original images of The Queen's College Middle Common Room (MCR), Oxford; the two images are taken from the same viewpoint, just rotating the camera (no parallax effect); **c** mosaic image; **d** mosaic with measurements superimposed.

and stitched to the first one to make the mosaic in Fig. 4.16**c**. If the homography between the original images and the world plane is known, then from the mosaicing process the homography between the final image and the world plane can be computed and measurements taken directly on the mosaic image as shown in Fig. 4.16**d**.

4.4 Duality and homologies

We have reserved the last section of this chapter to the investigation of some interesting properties of the plane rectification process and its relationship with homologies.

In an image of a planar surface in general position the plane appears projectively warped; the image plane, in contrast, is by definition front-on. By rectifying the image of the world plane via the estimated homography, a new image is obtained where the world plane is front-on but the original image plane is now projectively skewed; its boundary rectangle is warped into a projectively skewed quadrilateral (the solid quadrilateral around the image in Fig. 4.17**c**).

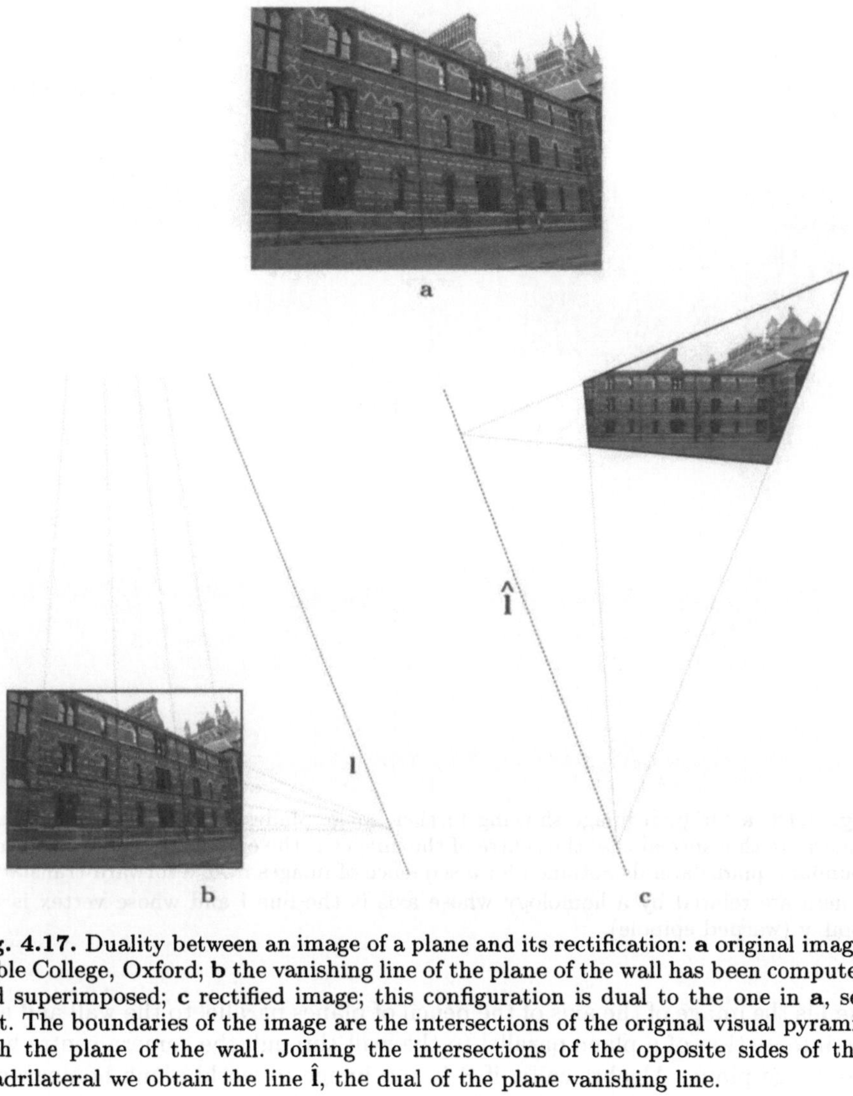

Fig. 4.17. Duality between an image of a plane and its rectification: **a** original image, Keble College, Oxford; **b** the vanishing line of the plane of the wall has been computed and superimposed; **c** rectified image; this configuration is dual to the one in **a**, see text. The boundaries of the image are the intersections of the original visual pyramid with the plane of the wall. Joining the intersections of the opposite sides of this quadrilateral we obtain the line $\hat{\mathbf{l}}$, the dual of the plane vanishing line.

The relationship between the image plane and the world plane is investigated via an example on a real scene.

Figure 4.17**b** shows the perspective image of a planar wall (seen before in Fig. 4.8**a**), the image plane is front-on (solid rectangle). Figure 4.17**c** is the image obtained by rectifying the wall plane. The edges of the image now make a non-rectangular quadrilateral. This is the perspectively distorted version of the rectangular boundary of the image in Fig. 4.17**a**; it can be interpreted as the intersection of the visual pyramid with the plane of the wall.

In Fig. 4.17**b** the vanishing line $\mathbf{l}$ for the wall plane has been computed by joining the intersections of two sets of parallel world edges lying on the wall. The

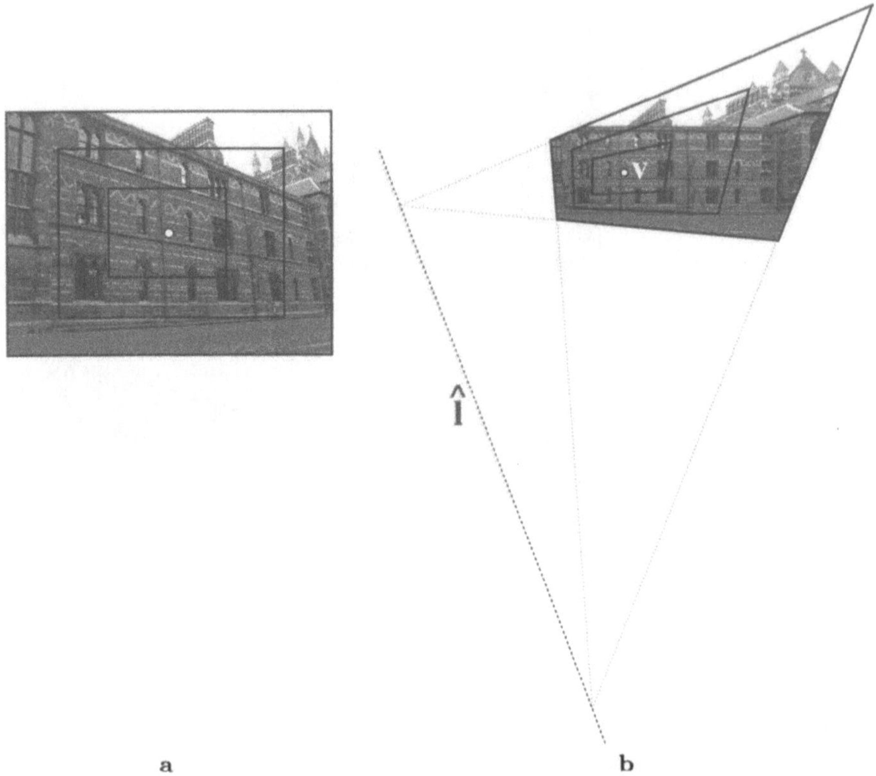

Fig. 4.18. **a** Original image showing further image planes for a forward translating camera. In this special case the centre of the images is the epipole. **b** The transformed boundary quadrilaterals obtained for a sequence of images from a forward translating camera are related by a homology whose axis is the line $\hat{\mathbf{l}}$ and whose vertex is the point $\mathbf{v}$ (warped epipole).

line $\mathbf{l}$ is the image of the axis of the pencil of planes parallel to the wall and also the intersection of a plane parallel to the wall through the camera centre with the image plane. Algebraically, if $\mathtt{H}$ is the image-to-world plane homography such that $\mathbf{X} = \mathtt{H}\mathbf{x}$, where $\mathbf{x}$ is an image point and $\mathbf{X}$ the corresponding world point, then $\mathbf{l}$ is the third row of the matrix $\mathtt{H}$.

On the other hand, the line $\hat{\mathbf{l}}$ in Fig. 4.17c joins the intersections of the opposite sides of the external quadrilateral. $\hat{\mathbf{l}}$ is the intersection of the *focal plane* with the plane of the wall and can be considered as the dual of the plane vanishing line $\mathbf{l}$. Algebraically $\hat{\mathbf{l}}$ is the third row of the matrix $\mathtt{H}^{-1}$.

Furthermore, section 3.1.3 has shown that images of parallel planes related by a parallel projection are related by a planar homology whose axis is the plane vanishing line and whose vertex is the vanishing point of the direction of projection. Dually we can imagine having a pencil of parallel image planes; this corresponds to having a sequence taken by a purely translating camera. By rectifying all the images in the sequence via the homographies induced by the world plane, we obtain the configuration illustrated in Fig. 4.18**b**. The corners of the warped image boundaries are related by a planar homology whose axis is the line $\hat{\mathbf{l}}$ and whose vertex is the rectified epipole. Figures 4.18**a** and **b** show an example of a special case where the camera is translating towards the college wall.

5. Single-view metrology

The previous chapter has investigated how measurements can be taken on planar surfaces from uncalibrated images. However, the world is not just one big plane (as Cristoforo Colombo discovered five centuries ago); it is a complex three-dimensional structure. Therefore, a more general analysis of the three-dimensional scene is required; this is achieved in this chapter. In particular, this chapter describes how aspects of the affine three-dimensional geometry of a scene may be measured from a single perspective image (see also [25, 26, 28]). The techniques described still concentrate on scenes containing planes and parallel lines, although the methods are not so restricted. The algorithms developed here extend and generalize previous results on single-view metrology [59, 68, 96, 100].

In this chapter we assume that the vanishing line of a *reference plane* in the scene may be determined from the image (see section 3.3), together with a vanishing point for a *reference direction* (not parallel to the plane). We are then concerned with three canonical types of measurement:

- measurements of the distance *between* any of the planes which are parallel to the reference plane;
- measurements *on* these planes (and comparison of these measurements with those obtained on any parallel plane);
- determining the camera's position in terms of the reference plane and direction.

The measurement methods developed here are independent of the camera internal parameters: focal length, aspect ratio, principal point and skew (uncalibrated camera, unknown internal parameters). We analyze situations where the projection matrix (external calibration) can only be partially determined from landmarks in the scene. This is an intermediate situation between calibrated reconstruction (where metric entities like angles between rays can be computed) and completely uncalibrated cameras (where a reconstruction can be obtained only up to a projective transformation).

The ideas in this chapter can be seen as reversing the rules for drawing perspective images given by Leon Battista Alberti [1] in his treatise on perspective (1435). These are the rules followed by the Italian Renaissance painters of the fifteenth century, and indeed the correctness of their mastery of perspective is demonstrated by analyzing the famous painting *Flagellazione* by Piero della Francesca (Fig. **4.9a**).

a b

Fig. 5.1. Measuring distances of points from a reference plane (the ground) in a single image: **a** the four pillars have the same height in the world, although their images clearly are not of the same length because of perspective effects; **b** as shown, however, all pillars are correctly measured to have the same height.

Section 5.1 gives simple geometric derivations of how, in principle, three-dimensional affine information may be extracted from one image (Fig. 5.1). Section 5.2 introduces an algebraic representation of the problem and shows that this representation unifies the three canonical measurement types, leading to simple formulae in each case. Section 5.3 describes how errors in image measurements propagate to errors in the three-dimensional measurements, and hence how to compute confidence intervals on the measurements, i.e. a quantitative assessment of accuracy. A generalization of the algorithm for recovering affine three-dimensional structure is presented in section 5.4. The work has a variety of applications, and three important ones are presented in section 5.5: forensic measurements, virtual modelling and furniture measurements.

5.1 Geometry

The camera model employed here is the central projection (described in section 3.1.1). We assume that the vanishing line of a reference plane in the scene may be computed from image measurements, together with a vanishing point for a direction not parallel to the plane (see section 3.3). Effects such as radial distortion, which corrupt the central projection model, can generally be removed as shown in section 3.2 and are therefore not detrimental to these methods. Although the schematic figures show the camera centre at a finite location, the results derived apply also to the case of a camera centre at infinity, i.e. where the images are obtained by parallel projection.

The basic geometry of the plane vanishing line and the vanishing point are illustrated in Fig. 5.2. The vanishing line $\mathbf{l}$ of the reference plane is the projection of the line at infinity of the reference plane into the image. The vanishing point $\mathbf{v}$ is the image of the point at infinity in the reference direction. Note that the reference direction need not be vertical, although for clarity we

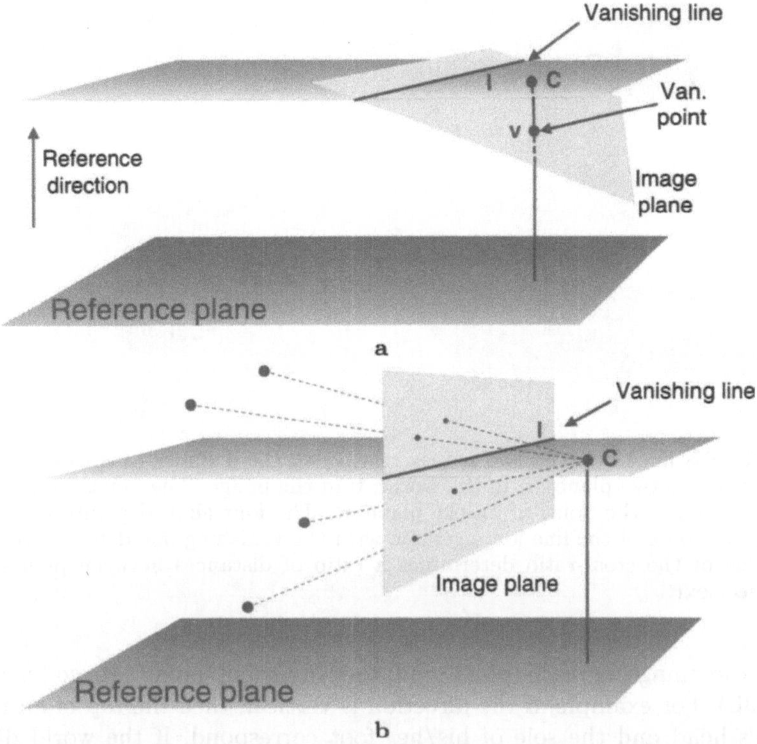

Fig. 5.2. Basic single-view geometry. **a** The plane's vanishing line **l** is the intersection of the image plane with a plane parallel to the reference plane and passing through the camera centre **C**. The vanishing point **v** is the intersection of the image plane with a line parallel to the reference direction through the camera centre. **b** The vanishing line partitions the space in the set of points closer to the reference plane than the camera and the set of points which are farther than the camera.

will often refer to the vanishing point as the "vertical" vanishing point. The vanishing point is then the image of the vertical "footprint" of the camera centre on the reference plane. Likewise, the reference plane will often, but not necessarily, be the ground plane, in which case the vanishing line is more commonly known as the "horizon".

It can be seen (for example, by inspection of Fig. 5.2) that the vanishing line partitions all points in the scene space. Any scene point which projects onto the vanishing line is at the same distance from the plane as the camera centre; if it lies "above" the line it is farther from the plane, and if "below" the vanishing line then it is closer to the plane than the camera centre.

We now give the following definition:

Definition 5.1.1. *Two points on separate planes (parallel to the reference plane) correspond if the line joining them is parallel to the reference direction.*

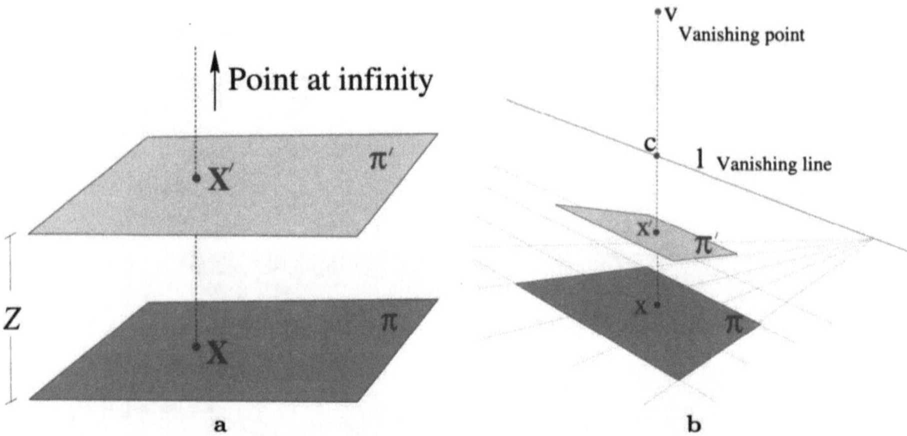

Fig. 5.3. Distance between two planes relative to the distance of the camera centre from one of the two planes: **a** in the world; **b** in the image. The point $\mathbf{x}$ on the plane $\boldsymbol{\pi}$ corresponds to the point $\mathbf{x}'$ on the plane $\boldsymbol{\pi}'$. The four aligned points $\mathbf{v}$, $\mathbf{x}$, $\mathbf{x}'$ and the intersection $\mathbf{c}$ of the line joining them with the vanishing line define a cross-ratio. The value of the cross-ratio determines a ratio of distances between planes in the world; see text.

Hence the image of each point and the vanishing point are collinear (see Fig. 5.3**b**). For example, if the direction is vertical, then the top of an upright person's head and the sole of his/her foot correspond. If the world distance between the two points is known, we term this a *reference distance.*

5.1.1 Measurements between parallel planes

We wish to measure the distance (in the reference direction) between two parallel planes, specified by the image points $\mathbf{x}$ and $\mathbf{x}'$. Figure 5.3 shows the geometry, with points $\mathbf{x}$ and $\mathbf{x}'$ in correspondence. The following theorem holds:

Theorem 5.1.1. *Given the vanishing line of a reference plane and the vanishing point for a reference direction, then distances from the reference plane parallel to the reference direction can be computed from their imaged end points up to a common scale factor. The scale factor can be determined from one known reference length.*

Proof. The four points $\mathbf{x}$, $\mathbf{x}'$, $\mathbf{c}$ and $\mathbf{v}$ marked on Fig. 5.3**b** define a cross-ratio [117]. The vanishing point is the image of a point at infinity in the scene and the point $\mathbf{c}$, since it lies on the vanishing line, is the image of a point at distance Z_c from the plane $\boldsymbol{\pi}$, where Z_c is the distance of the camera centre from $\boldsymbol{\pi}$. In the world, the value of the cross-ratio provides an affine length ratio, which determines the distance Z between the planes containing $\mathbf{X}'$ and $\mathbf{X}$ (in Fig. 5.3**a**) relative to the camera's distance Z_c from the plane $\boldsymbol{\pi}$ (or $\boldsymbol{\pi}'$ depending on the ordering of the cross-ratio). Note that the distance Z

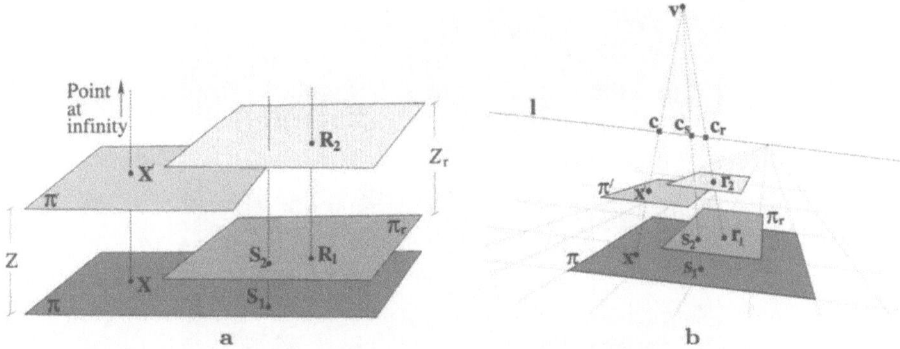

Fig. 5.4. Distance between two planes relative to the distance between two other planes: **a** in the world; **b** in the image. The point $\mathbf{x}$ on the plane π corresponds to the point $\mathbf{x}'$ on the plane π'. The point $\mathbf{s}_1$ corresponds to the point $\mathbf{s}_2$. The point $\mathbf{r}_1$ corresponds to the point $\mathbf{r}_2$. The distance Z_r in the world between $\mathbf{R}_1$ and $\mathbf{R}_2$ is known and used as reference to compute the distance Z; see text.

can alternatively be computed using a line-to-line homography avoiding the ordering ambiguity of the cross-ratio. For the case in Fig. 5.3**b** we obtain:

$$\frac{Z}{Z_c} = 1 - \frac{d(\mathbf{x}', \mathbf{c})\, d(\mathbf{x}, \mathbf{v})}{d(\mathbf{x}, \mathbf{c})\, d(\mathbf{x}', \mathbf{v})} \tag{5.1}$$

where $d(\mathbf{x}_1, \mathbf{x}_2)$ is the distance between two generic image points $\mathbf{x}_1$ and $\mathbf{x}_2$.

The absolute distance Z can be obtained from the distance ratio (5.1) once the camera's distance Z_c is specified. However it is usually more practical to determine the distance Z via a second measurement in the image, that of a known reference length. In fact, given a known reference distance Z_r, the distance of the camera Z_c can be computed from (5.1) and then the distance Z determined by applying (5.1) to a new pair of end points. □

Theorem 5.1.1 can be generalized to the following:

Theorem 5.1.2. *Given a set of linked parallel planes, the distance between any pair of planes is sufficient to determine the absolute distance between any other pair. The link is provided by a chain of point correspondences between the set of planes.*

Proof. Figure 5.4 shows a diagram where four parallel planes are imaged. They all share the same vanishing line which is the image of the axis of the pencil. The distance Z_r between two of them can be used as reference to compute the distance Z between the other two as follows:

- from the cross-ratio defined by the four aligned points $\mathbf{v}$, $\mathbf{c}_r$, $\mathbf{r}_2$, $\mathbf{r}_1$ and the known distance Z_r between the world points $\mathbf{R}_1$ and $\mathbf{R}_2$, the distance of the camera from the plane π_r can be computed;
- that camera distance and the cross-ratio defined by the four aligned points $\mathbf{v}$, $\mathbf{c}_s$, $\mathbf{s}_2$, $\mathbf{s}_1$, determine the distance between the planes π_r and π. The distance Z_c of the camera from the plane π is therefore determined too;

Fig. 5.5. Measuring the height of a person from a single image: **a** original image; **b** the height of the person is computed from the image as 178.8cm; the true height is 180 cm, but note that the person is leaning down a bit on his right foot. The vanishing line is shown in white; the vertical vanishing point is not shown since it lies well below the image. The reference distance is in white (the height of the window frame on the right). Compare the marked points with the ones in Fig. 5.4.

- the distance Z_c can now be used in (5.1) to compute the distance Z between the two planes $\boldsymbol{\pi}$ and $\boldsymbol{\pi}'$.

□

Section 5.2.1 presents an algebraic derivation of these results which avoids the need to compute the distance of the camera explicitly and simplifies the measurement procedure.

Example.
Figure 5.5 shows that a person's height may be computed from an image given a vertical reference distance elsewhere in the scene. The ground plane is reference. The height of the frame of the window has been measured on site and used as the reference distance (it corresponds to the distance between $\mathbf{R}_1$ and $\mathbf{R}_2$ in the world in Fig. 5.4**a**). This situation corresponds to the one in Fig. 5.4 where the two points $\mathbf{S}_2$ and $\mathbf{R}_1$ (and therefore $\mathbf{s}_2$ and $\mathbf{r}_1$) coincide. The height of the person is computed from the cross-ratio defined by the points $\mathbf{x}'$, $\mathbf{c}$, $\mathbf{x}$ and the

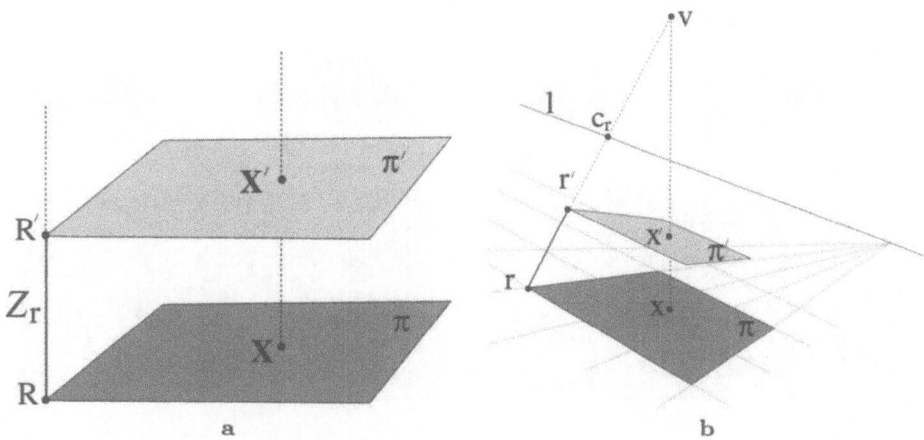

Fig. 5.6. Homology mapping between imaged parallel planes related by parallel projection. **a** In the world a point $\mathbf{X}$ on the plane $\boldsymbol{\pi}$ is mapped into the point $\mathbf{X}'$ on $\boldsymbol{\pi}'$ by a parallel projection. **b** In the image the mapping between the images of the two planes is a homology, where $\mathbf{v}$ is the *vertex* and $\mathbf{l}$ the *axis*. The correspondence $\mathbf{r} \rightarrow \mathbf{r}'$ fixes the remaining degree of freedom of the homology from the cross-ratio of the four points $\mathbf{v}$, $\mathbf{c}_r$, $\mathbf{r}'$ and $\mathbf{r}$.

vanishing point (Fig. 5.4**b**) as described in the proof above. Since the points $\mathbf{S}_2$ and $\mathbf{R}_1$ coincide, the derivation is simpler.

5.1.2 Measurements on parallel planes

If the reference plane $\boldsymbol{\pi}$ is affine calibrated (its vanishing line is known), then from image measurements we can compute:

1. ratios of lengths of parallel line segments on the plane;
2. ratios of areas on the plane.

Moreover, the vanishing line is shared by the pencil of planes parallel to the reference plane, hence affine measurements may be obtained for any other plane in the pencil. However, although affine measurements, such as an area ratio, may be made *on* a particular plane, the areas of regions lying on two parallel planes cannot be compared directly. If the region is parallel projected in the scene from one plane onto the other, affine measurements can then be made from the image since both regions are now on the same plane, and parallel projection between parallel planes does not alter affine properties.

A map in the world between parallel planes induces a map in the image between images of points on the two planes. As stated in section 3.1.3, this image map is a *planar homology*. The geometry is illustrated in Fig. 5.6. In this case the vanishing line of the plane, and the vertical vanishing point, are, respectively, the *axis* and *vertex* of the homology which relates a pair of planes in the pencil.

The homology $\tilde{\mathtt{H}}$ can then be parametrized as in (3.5):

a b

Fig. 5.7. Measuring the ratio of lengths of parallel line segments lying on two parallel scene planes. **a** Original image, the facade of a building. The two front-facing planes are parallel. We want to compare lengths of parallel segments on the two planes. **b** The points $\mathbf{r}$ and $\mathbf{r}'$ (together with the plane vanishing line and the vanishing point) define the homology between the two front-facing planes.

$$\tilde{\mathtt{H}} = \mathtt{I} + \mu \frac{\mathbf{v}\mathbf{l}^\top}{\mathbf{v} \cdot \mathbf{l}} \tag{5.2}$$

where $\mathbf{v}$ is the vanishing point, $\mathbf{l}$ is the plane vanishing line and μ is the characteristic ratio. Thus $\mathbf{v}$ and $\mathbf{l}$ specify four of the five degrees of freedom of the homology. The remaining degree of freedom of the homology, μ, is uniquely determined from any pair of image points which correspond between the planes (points $\mathbf{r}$ and $\mathbf{r}'$ in figure 5.6).

Once the matrix $\tilde{\mathtt{H}}$ is computed each point on a plane can be transferred into the corresponding point on a parallel plane as $\mathbf{x}' = \tilde{\mathtt{H}}\mathbf{x}$. An example of this *homology mapping* is shown in figure 3.7. Consequently measurements made on two separate planes can be compared. In particular, we may compute:

1. the ratio between two parallel lengths, one length on each plane;
2. the ratio between two areas, one area on each plane.

In fact all points can be transferred from one plane to the reference plane using the homology. Since the reference plane's vanishing line is known, affine measurements in the plane can be made (e.g. parallel length or area ratios).

Example.
Figure 5.7 shows an example. The vanishing line of the two front-facing walls and the vanishing point are known, as is the point correspondence $\mathbf{r}$, $\mathbf{r}'$ in the reference direction. The ratio of the length of parallel line segments is computed by using the formulae given in section 5.2.2. Notice that errors in the selection of point positions affect the computations; the veridical values of the ratios in Fig. 5.7**b** are exact integers. A proper error analysis is necessary to estimate the uncertainty of these affine measurements.

5.1.3 Determining the camera position

Section 5.1.1 computes the distances between planes as a ratio relative to the camera's distance from the reference plane. Conversely, the camera's distance Z_c from a particular plane can be obtained knowing a single reference distance Z_r.

Furthermore, Fig. 5.2 shows that the location of the camera relative to the reference plane is the back-projection of the vertical vanishing point onto the reference plane. This back-projection is accomplished by a homography which maps the image to the reference plane (and vice versa). Although the choice of the world coordinate frame is somewhat arbitrary, fixing this frame immediately defines the homography uniquely and hence the complete camera position.

5.2 Algebraic representation

The measurements described in the previous section are computed in terms of cross-ratios. This section develops a uniform algebraic approach to the problem which has a number of advantages over direct geometric construction: (i) it avoids potential problems with ordering for the cross-ratio; (ii) it allows to deal with both minimal or over-constrained configurations uniformly; (iii) different types of measurement are unified within one representation; (iv) in section 5.3 this algebraic representation is used to develop an uncertainty analysis for measurements.

To begin, an affine coordinate system XYZ is defined in space [69, 98]. Let the origin of the coordinate frame lie on the reference plane, with the X- and Y-axes spanning the plane. The Z-axis is the reference direction, which is thus any direction not parallel to the plane. The image coordinate system is the usual xy affine image frame, and a point $\mathbf{X}$ in space is projected to the image point $\mathbf{x}$ via a 3×4 projection matrix $\mathtt{P}$ as:

$$\mathbf{x} = \mathtt{P}\mathbf{X} = \begin{bmatrix} \mathbf{p}_1 & \mathbf{p}_2 & \mathbf{p}_3 & \mathbf{p}_4 \end{bmatrix} \mathbf{X}$$

where $\mathbf{x}$ and $\mathbf{X}$ are homogeneous vectors in the form: $\mathbf{x} = (x, y, w)^\top$, $\mathbf{X} = (X, Y, Z, W)^\top$, and "=" means equality up to scale.

Denoting the vanishing points for the X, Y and Z directions as (respectively) $\mathbf{v}_X$, $\mathbf{v}_Y$ and $\mathbf{v}$, it is clear by inspection [38] that the first three columns of $\mathtt{P}$ are the vanishing points: $\mathbf{v}_X = \mathbf{p}_1$, $\mathbf{v}_Y = \mathbf{p}_2$ and $\mathbf{v} = \mathbf{p}_3$; and that the final column of $\mathtt{P}$ is the projection of the origin of the world coordinate system: $\mathbf{o} = \mathbf{p}_4$. Since our choice of coordinate frame has the X and Y axes in the reference plane $\mathbf{p}_1 = \mathbf{v}_X$ and $\mathbf{p}_2 = \mathbf{v}_Y$ are two distinct points on the vanishing line. Choosing these fixes the X and Y affine coordinate axes. The vanishing line is denoted by $\mathbf{l}$, and to emphasize that the vanishing points $\mathbf{v}_X$ and $\mathbf{v}_Y$ lie on it, we denote them by $\mathbf{l}_1^\perp$, $\mathbf{l}_2^\perp$, with $\mathbf{l}_i^\perp \cdot \mathbf{l} = 0$.

Columns 1, 2 and 4 of the projection matrix are the three columns of the reference plane to image homography. This homography must have rank three, otherwise the reference plane to image map is degenerate. Consequently, the final column (the origin of the coordinate system) must not lie on the vanishing line, since if it does then all three columns are points on the vanishing line, and thus are not linearly independent. Hence we set it to be $\mathbf{o} = \mathbf{p}_4 = \mathbf{l}/||\mathbf{l}|| = \bar{\mathbf{l}}$.

Therefore the final parameterization of the projection matrix P is:

$$\mathtt{P} = \left[\ \mathbf{l}_1^{\perp}\ \mathbf{l}_2^{\perp}\ \alpha\mathbf{v}\ \bar{\mathbf{l}}\ \right] \tag{5.3}$$

where α is a scale factor, which has an important role to play in the remainder of the chapter.

Note that the vertical vanishing point $\mathbf{v}$ imposes two constraints on the P matrix, the vanishing line $\mathbf{l}$ two constraints also and the α parameter only one for a total of five independent constraints. In general, however, the P matrix has eleven degrees of freedom, which can be regarded as comprising eight for the world-to-image homography induced by the reference plane, two for the vanishing point and one for the affine parameter α. In the present case the vanishing line determines two of the eight degrees of freedom of the homography.

The following sections show how to compute various measurements from this projection matrix. Measurements of distances between planes are independent of the first two (in general under-determined) columns of P. If $\mathbf{v}$ and $\mathbf{l}$ are specified, the only unknown quantity for these measurements is α. Coordinate measurements within the planes depend on the first two and the fourth columns of P. They define an affine coordinate frame within the plane. Affine measurements (e.g. area ratios), though, are independent of the actual coordinate frame and depend only on the fourth column of P. If any metric information on the plane is known, that can be used to impose constraints on the choice of the frame.

5.2.1 Measurements between parallel planes

Distance of a plane from the reference plane π.

We wish to measure the distance between scene planes specified by a point $\mathbf{X}$ and a point $\mathbf{X}'$ in the scene (see Fig. 5.3a). These points may be chosen as respectively $\mathbf{X} = (X, Y, 0)^\top$ and $\mathbf{X}' = (X, Y, Z)^\top$, and their images are $\mathbf{x}$ and $\mathbf{x}'$ (Fig. 5.8). If P is the projection matrix, then the image coordinates are:

$$\mathbf{x} = \mathtt{P}\begin{pmatrix} X \\ Y \\ 0 \\ 1 \end{pmatrix}, \qquad \mathbf{x}' = \mathtt{P}\begin{pmatrix} X \\ Y \\ Z \\ 1 \end{pmatrix}$$

The equations above can be rewritten as:

$$\mathbf{x} = \rho(X\mathbf{p}_1 + Y\mathbf{p}_2 + \mathbf{p}_4) \tag{5.4}$$

$$\mathbf{x}' = \rho'(X\mathbf{p}_1 + Y\mathbf{p}_2 + Z\mathbf{p}_3 + \mathbf{p}_4) \tag{5.5}$$

where ρ and ρ' are unknown scale factors, and $\mathbf{p}_i$ is the ith column of the P matrix.

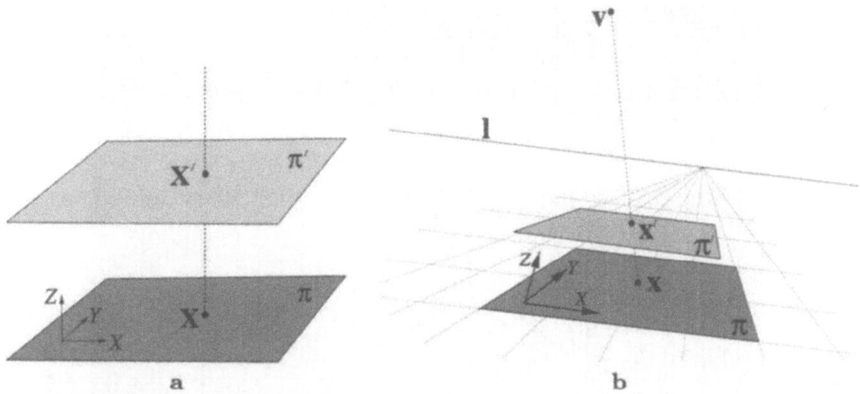

Fig. 5.8. Measuring the distance of a plane $\boldsymbol{\pi}'$ from the parallel reference plane $\boldsymbol{\pi}$: **a** in the world; **b** in the image.

Since $\mathbf{p}_1 \cdot \bar{\mathbf{l}} = \mathbf{p}_2 \cdot \bar{\mathbf{l}} = 0$ and $\mathbf{p}_4 \cdot \bar{\mathbf{l}} = 1$, taking the scalar product of (5.4) with $\bar{\mathbf{l}}$ yields $\rho = \bar{\mathbf{l}} \cdot \mathbf{x}$ and therefore (5.5) can be rewritten as:

$$\mathbf{x}' = \rho' \left(\frac{\mathbf{x}}{\rho} + \alpha Z \mathbf{v} \right) \tag{5.6}$$

By taking the vector product of both terms of (5.6) with $\mathbf{x}'$ we obtain

$$\mathbf{x} \times \mathbf{x}' = -\alpha Z \rho (\mathbf{v} \times \mathbf{x}') \tag{5.7}$$

and, finally, taking the norm of both sides of (5.7) yields:

$$\alpha Z = -\frac{||\mathbf{x} \times \mathbf{x}'||}{(\bar{\mathbf{l}} \cdot \mathbf{x})||\mathbf{v} \times \mathbf{x}'||} \tag{5.8}$$

Since αZ scales linearly with α, we have obtained affine structure. If α is known, then we immediately obtain a metric value for Z as:

$$Z = -\frac{||\mathbf{x} \times \mathbf{x}'||}{(\mathbf{p}_4 \cdot \mathbf{x})||\mathbf{p}_3 \times \mathbf{x}'||} \tag{5.9}$$

Conversely, if Z is known (i.e. it is a reference distance) then (5.8) provides a means of computing α, and hence removing the affine ambiguity.

Metric calibration from multiple references.
If more than one reference distance is known, then an estimate of α can be derived from an error minimization algorithm. Here a special case is considered where all distances are measured from the same reference plane and an algebraic error is minimized. An optimal minimization algorithm will be described in section 5.3.2.

For the ith reference distance Z_i with end points $\mathbf{r}_i$ and $\mathbf{r}'_i$, we define: $\beta_i = ||\mathbf{r}_i \times \mathbf{r}'_i||$, $\rho_i = \bar{\mathbf{l}} \cdot \mathbf{r}_i$, $\gamma_i = ||\mathbf{v} \times \mathbf{r}'_i||$. Therefore, from (5.8):

$$\alpha Z \rho_i \gamma_i = -\beta_i \tag{5.10}$$

Fig. 5.9. Measuring heights using parallel lines: The vertical vanishing point and the vanishing line for the ground plane have been computed. The distance of the top of the window on the left wall from the ground is known and used as reference. The distance of the top of the window on the right wall from the ground is computed from the distance between the two horizontal lines whose images are $\mathbf{l}_{x'}$ and $\mathbf{l}_x$. The top line $\mathbf{l}_{x'}$ is defined by the top edge of the window, and the line $\mathbf{l}_x$ is the corresponding one on the ground plane. The distance between them is computed to be 294.3 cm.

Note that all the points $\mathbf{r}_i$ are images of world points $\mathbf{R}_i$ on the reference plane π. By reorganizing (5.10) the following matrix $\mathtt{A}$ is defined:

$$\mathtt{A} = \begin{pmatrix} Z_1\rho_1\gamma_1 & \beta_1 \\ \vdots & \vdots \\ Z_i\rho_i\gamma_i & \beta_i \\ \vdots & \vdots \\ Z_n\rho_n\gamma_n & \beta_n \end{pmatrix}$$

where n is the number of reference distances.

If the references are not affected by measurement error or $n = 1$, then $\mathtt{A}\mathbf{s} = \mathbf{0}$ where $\mathbf{s} = (s_1 \; s_2)^\top$ is a homogeneous 2-vector and

$$\alpha = \frac{s_1}{s_2} \tag{5.11}$$

In general, $n > 1$ and uncertainty is present in the reference distances. In this case we find the solution $\mathbf{s}$ which minimizes $||\mathtt{A}\mathbf{s}||$. That is the eigenvector of the 2×2 matrix $\mathtt{M} = \mathtt{A}^\top\mathtt{A}$ corresponding to its minimum eigenvalue. The parameter α is finally computed from (5.11). With more reference distances Z_i, α is estimated more accurately (see section 5.3), but no more constraints are added on the $\mathtt{P}$ matrix.

– Worked Example –

Objective: In Fig. 5.9 the distance of a horizontal line from the ground is measured.

- The vertical vanishing point $\mathbf{v}$ is computed by intersecting vertical (scene) edges.

All images of lines parallel to the ground plane intersect in points on the horizon, therefore:

- a vanishing point $\mathbf{v}_1$ on the horizon is computed by intersecting the edges of the planks on the right side of the shed;
- a second vanishing point $\mathbf{v}_2$ is computed by intersecting the edges of the planks on the left side of the shed and the parallel edges on the roof;
- the plane vanishing line $\mathbf{l}$ is computed by joining those two points: $\mathbf{l} = \mathbf{v}_1 \times \mathbf{v}_2$;
- the distance of the top of the frame of the window on the left from the ground has been measured on site and used as reference to compute α, as in (5.8).
- the line $\mathbf{l}_{x'}$, the image of a horizontal line, is selected in the image by choosing any two points on it;
- the associated vanishing point $\mathbf{v}_h$ is computed as $\mathbf{v}_h = \mathbf{l}_{x'} \times \mathbf{l}$;
- the line $\mathbf{l}_x$, which is the image of a line parallel to $\mathbf{l}_{x'}$ in the scene is constrained to pass through $\mathbf{v}_h$, therefore $\mathbf{l}_x$ is specified by choosing one additional point on it;
- a point $\mathbf{x}'$ is selected along the line $\mathbf{l}_{x'}$ and its corresponding point $\mathbf{x}$ on the line $\mathbf{l}_x$ computed as $\mathbf{x} = (\mathbf{x}' \times \mathbf{v}) \times \mathbf{l}_x$;
- equation (5.9) is now applied to the pair of points $\mathbf{x}$, $\mathbf{x}'$ to compute the distance $Z = 294.3$ cm.

Distance between any two parallel planes.

The projection matrix $\mathtt{P}$ from the world to the image is defined in (5.3) with respect to a coordinate frame on the reference plane (Fig. 5.8). In this section we determine the projection matrix $\mathtt{P}'$ referred to the parallel plane $\boldsymbol{\pi}'$ and we show how distances from the plane $\boldsymbol{\pi}'$ can be computed.

Suppose the world coordinate system is translated by Z_r from the plane $\boldsymbol{\pi}$ onto the plane $\boldsymbol{\pi}'$ along the reference direction (Fig. 5.10), then the new projection matrix $\mathtt{P}'$ can be parametrized as:

$$\mathtt{P}' = \left[\, \mathbf{p}_1 \; \mathbf{p}_2 \; \mathbf{p}_3 \; Z_r\mathbf{p}_3 + \mathbf{p}_4 \,\right] \tag{5.12}$$

Note that only the last column has changed, and if $Z_r = 0$ then $\mathtt{P}' = \mathtt{P}$, as expected.

The distance Z' of the plane $\boldsymbol{\pi}''$ from the plane $\boldsymbol{\pi}'$ in space can be computed as (cf. (5.9)).

$$Z' = -\frac{\|\mathbf{x}' \times \mathbf{x}''\|}{\rho'\|\mathbf{p}_3 \times \mathbf{x}''\|} \quad \text{with} \quad \rho' = \frac{\mathbf{p}_4 \cdot \mathbf{x}'}{1 + Z_r\mathbf{p}_3 \cdot \mathbf{p}_4} \tag{5.13}$$

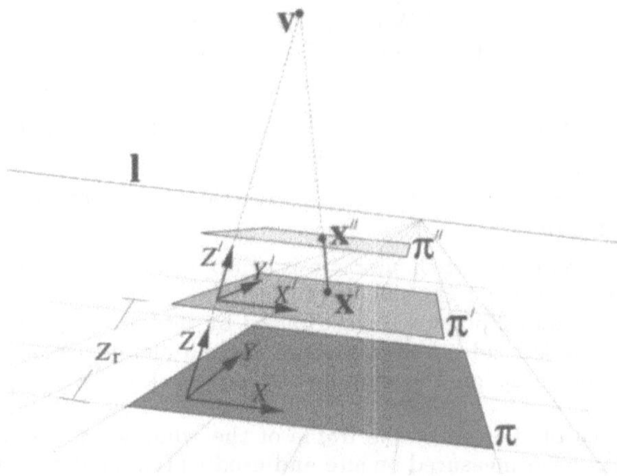

Fig. 5.10. Measuring the distance between any two planes $\boldsymbol{\pi}'$ and $\boldsymbol{\pi}''$ parallel to the reference plane $\boldsymbol{\pi}$: the geometry.

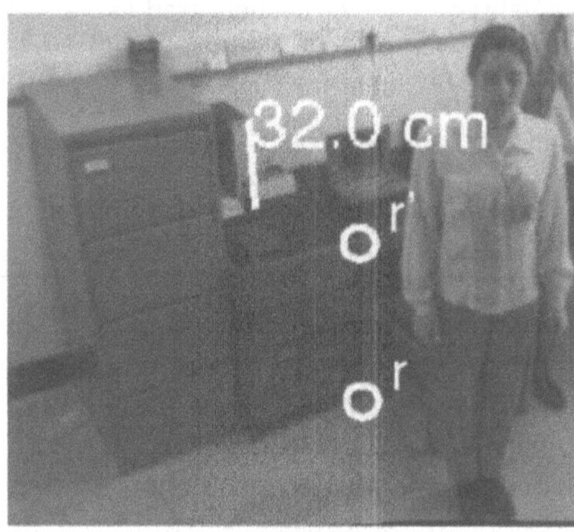

Fig. 5.11. Measuring heights of objects on separate planes: the height of the desk is known and the height of the file on the desk is computed.

– Worked Example –

Objective: In Fig. 5.11 the height of a file on a desk is computed from the height of the desk itself:

- the ground is the reference plane $\boldsymbol{\pi}$ and the top of the desk is the plane denoted as $\boldsymbol{\pi}'$ in Fig. 5.10;
- the plane vanishing line and vertical vanishing point are computed as usual by intersecting parallel edges;
- the distance Z_r between the points $\mathbf{r}$ and $\mathbf{r}'$ is known (the height of the desk has been measured on site) and used to compute the α parameter from (5.8);
- equation (5.13) is now applied to the end points of the marked segment to compute the height $Z' = 32.0$ cm.

5.2.2 Measurements on parallel planes

As described in section 5.1.2, given the homology between two planes $\boldsymbol{\pi}$ and $\boldsymbol{\pi}'$ in the pencil, we can transfer all points from one plane to the other and make affine measurements in either plane.

The homology between the planes can be derived directly from the two projection matrices (5.3) and (5.12). The plane-to-image homographies may be extracted from the projection matrices $\mathtt{P}$ and $\mathtt{P}'$, ignoring the third column, to give:

$$\mathtt{H} = \left[\ \mathbf{p}_1 \ \mathbf{p}_2 \ \mathbf{p}_4 \ \right], \quad \mathtt{H}' = \left[\ \mathbf{p}_1 \ \mathbf{p}_2 \ Z_r\mathbf{p}_3 + \mathbf{p}_4 \ \right]$$

Then $\tilde{\mathtt{H}} = \mathtt{H}'\mathtt{H}^{-1}$ maps image points on the plane $\boldsymbol{\pi}$ onto points on the plane $\boldsymbol{\pi}'$ and so defines the homology. By inspection, since $\mathbf{p}_1 \cdot \mathbf{p}_4 = 0$ and $\mathbf{p}_2 \cdot \mathbf{p}_4 = 0$ then $(\mathtt{I} + Z_r\mathbf{p}_3\mathbf{p}_4^\top)\mathtt{H} = \mathtt{H}'$, hence the homology matrix $\tilde{\mathtt{H}}$ is:

$$\tilde{\mathtt{H}} = \mathtt{I} + Z_r\mathbf{p}_3\mathbf{p}_4^\top \tag{5.14}$$

Alternatively from (5.3) the homology matrix can be written as:

$$\tilde{\mathtt{H}} = \mathtt{I} + \psi\mathbf{v}\bar{\mathbf{l}}^\top \tag{5.15}$$

with $\psi = \alpha Z_r$.

If the distance Z_r and the last two columns of the matrix $\mathtt{P}$ are known then the homology between the two planes $\boldsymbol{\pi}$ and $\boldsymbol{\pi}'$ is computed as in (5.14). Otherwise, if only $\mathbf{v}$ and $\mathbf{l}$ are known and two corresponding points $\mathbf{r}$ and $\mathbf{r}'$ are viewed, then the homology characteristic parameter ψ in (5.15) can be computed from (5.8). In fact:

$$\alpha Z_r = \psi = -\frac{||\mathbf{r} \times \mathbf{r}'||}{(\bar{\mathbf{l}} \cdot \mathbf{r})||\mathbf{v} \times \mathbf{r}'||} \tag{5.16}$$

knowing either the distance Z_r between the two planes or the α parameter.

Examples of homology transfer and affine measurements are shown in Figs 5.7 and 5.12.

– Worked Example –

Objective: In Fig. 5.12 the ratio between the areas of two windows $\frac{A_1}{A_2}$ in the world is computed:

- the orthogonal vanishing point $\mathbf{v}$ is computed by intersecting the edges of the small windows linking the two front planes;
- the plane vanishing line $\mathbf{l}$ (common to both front planes) is computed by intersecting two sets of parallel edges on the two planes;
- the only remaining parameter ψ of the homology $\tilde{\mathtt{H}}$ in (5.15) is computed from (5.16);
- each of the four corners of the window on the left is transferred by the homology $\tilde{\mathtt{H}}$ onto the corresponding points on the plane of the other window (Fig. 5.12**b**);

Now we have two quadrilaterals on the same plane:

- the image is affine-warped by pulling the plane vanishing line to infinity [78];
- the ratio between the two areas in the world is computed as the ratio between the areas in the affine-warped image; we obtain $\frac{A_1}{A_2} = 1.45$.

Fig. 5.12. Measuring ratios of areas on separate planes: **a** original image with two windows highlighted; **b** the left window is transferred onto the plane identified by $\mathbf{r}'$ by the homology mapping (5.15). The two areas now lie on the same plane and can, therefore, be compared. The ratio between the areas of the two windows is then computed as: $\frac{A_1}{A_2} = 1.45$.

5.2.3 Determining the camera position

Suppose the camera centre is $\mathbf{C} = (X_c, Y_c, Z_c, W_c)^\top$ in affine coordinates (see Fig. 5.2). Then since $\mathtt{P}\mathbf{C} = \mathbf{0}$ we have

$$\mathtt{P}\mathbf{C} = \mathbf{p}_1 X_c + \mathbf{p}_2 Y_c + \mathbf{p}_3 Z_c + \mathbf{p}_4 W_c = \mathbf{0} \tag{5.17}$$

The solution to this set of equations is given (using Cramer's rule) by:

$$X_c = -\det\left[\,\mathbf{p}_2\ \mathbf{p}_3\ \mathbf{p}_4\,\right], \quad Y_c = \det\left[\,\mathbf{p}_1\ \mathbf{p}_3\ \mathbf{p}_4\,\right],$$
$$Z_c = -\det\left[\,\mathbf{p}_1\ \mathbf{p}_2\ \mathbf{p}_4\,\right], \quad W_c = \det\left[\,\mathbf{p}_1\ \mathbf{p}_2\ \mathbf{p}_3\,\right] \tag{5.18}$$

If $\mathtt{P}$ is entirely known the location of the camera centre is metric defined. If α is unknown we can write:

$$X_c = -\det\left[\,\mathbf{p}_2\ \mathbf{v}\ \mathbf{p}_4\,\right], \quad Y_c = \det\left[\,\mathbf{p}_1\ \mathbf{v}\ \mathbf{p}_4\,\right],$$
$$\alpha Z_c = -\det\left[\,\mathbf{p}_1\ \mathbf{p}_2\ \mathbf{p}_4\,\right], \quad W_c = \det\left[\,\mathbf{p}_1\ \mathbf{p}_2\ \mathbf{v}\,\right] \tag{5.19}$$

and obtain the distance Z_c of the camera centre from the plane up to the affine scale factor α. As before, the distance Z_c can be upgraded to metric with knowledge of α, or use knowledge of the camera height to compute α and upgrade the affine structure.

Note that affine viewing conditions (where the camera centre is at infinity) present no problem in expressions (5.18) and (5.19), since in this case we have $\bar{\mathbf{l}} = \left[0\,0\,*\right]^\top$ and $\mathbf{v} = \left[*\,*\,0\right]^\top$. Hence $W_c = 0$, thus a camera centre on the plane at infinity is obtained, as expected. This point on $\boldsymbol{\pi}_\infty$ represents the viewing direction for the parallel projection.

Fig. 5.13. Computing the location of the camera: equations (5.18) are used to obtain: X_c = -381.0 cm, Y_c = -653.7 cm, Z_c = 162.8 cm.

If the viewpoint is finite (i.e. not affine viewing conditions) then the formula for αZ_c may be developed further by taking the scalar product of both sides of (5.17) with the vanishing line $\bar{\mathbf{l}}$. The result is:

$$\alpha Z_c = -(\bar{\mathbf{l}} \cdot \mathbf{v})^{-1} \tag{5.20}$$

– Worked Example –

Objective: In Fig. 5.13 the position of the camera centre with respect to the chosen Cartesian coordinate system is determined. Note that in this case we have chosen $\mathbf{p}_4$ to be the point $\mathbf{o}$ in the figure instead of $\bar{\mathbf{l}}$:

- the ground plane (X, Y plane) is the reference;
- the vertical vanishing point is computed by intersecting vertical edges;
- the two sides of the rectangular base of the porch have been measured, thus providing the position of four points on the reference plane; the world-to-image homography is computed from those points [24];
- the distance of the top of the frame of the window on the left from the ground has been measured on site and used as reference to compute α, as in (5.8);
- the three-dimensional position of the camera centre is then computed simply by applying equations (5.18); we obtain

$$X_c = -381.0\text{cm} \quad Y_c = -653.7\text{cm} \quad Z_c = 162.8\text{cm}$$

In Fig. 5.23**c**, the camera has been superimposed into a virtual view of the reconstructed scene.

5.3 Uncertainty analysis

Feature detection and extraction, whether manual or automatic (e.g. using an edge detector), can only be achieved to a finite accuracy. Any features extracted from an image, therefore, are subject to measurements errors. In this section we consider how these errors propagate through the measurement formulae in order to quantify the uncertainty of the final measurements [38]. This is achieved by using a first order error analysis (see section 3.4).

The uncertainty of the projection matrix is analyzed first, followed by the uncertainty of the distance measurements.

5.3.1 Uncertainty of the P matrix

The uncertainty in the projection matrix depends on the location of the vanishing line, the location of the vanishing point, and on α, the affine scale factor. Since only the final two columns contribute, the uncertainty in P is modelled as a 6×6 homogeneous covariance matrix, $\Lambda_{\mathtt{P}}$. Since the two columns have only five degrees of freedom (two for $\mathbf{v}$, two for $\mathbf{l}$ and one for α), the covariance matrix is singular, with rank five.

Assuming statistical independence between the two column vectors $\mathbf{p}_3$ and $\mathbf{p}_4$, the 6×6 rank-five covariance matrix $\Lambda_{\mathtt{P}}$ can be written as:

$$\Lambda_{\mathtt{P}} = \begin{pmatrix} \Lambda_{\mathbf{p}_3} & 0 \\ 0 & \Lambda_{\mathbf{p}_4} \end{pmatrix} \tag{5.21}$$

Furthermore, assuming statistical independence between α and $\mathbf{v}$, since $\mathbf{p}_3 = \alpha\mathbf{v}$:

$$\Lambda_{\mathbf{p}_3} = \alpha^2 \Lambda_{\mathbf{v}} + \sigma_\alpha^2 \mathbf{v}\mathbf{v}^\top \tag{5.22}$$

with $\Lambda_{\mathbf{v}}$ the homogeneous 3×3 covariance of the vanishing point $\mathbf{v}$, and the variance σ_α^2 computed as in Appendix D.

Since $\mathbf{p}_4 = \bar{\mathbf{l}} = \frac{\mathbf{l}}{\|\mathbf{l}\|}$ its covariance is:

$$\Lambda_{\mathbf{p}_4} = \frac{\partial \mathbf{p}_4}{\partial \mathbf{l}} \Lambda_{\mathbf{l}} \frac{\partial \mathbf{p}_4}{\partial \mathbf{l}}^\top \tag{5.23}$$

where the 3×3 Jacobian $\frac{\partial \mathbf{p}_4}{\partial \mathbf{l}}$ is:

$$\frac{\partial \mathbf{p}_4}{\partial \mathbf{l}} = \frac{\mathbf{l} \cdot \mathbf{l}\mathtt{I} - \mathbf{l}\mathbf{l}^\top}{(\mathbf{l} \cdot \mathbf{l})^{\frac{3}{2}}}$$

5.3.2 Uncertainty of measurements between planes

When making measurements between planes (5.9), uncertainty arises from the uncertain image locations of the points $\mathbf{x}$ and $\mathbf{x}'$ and from the uncertainty in P.

The uncertainty in the end points $\mathbf{x}, \mathbf{x}'$ of the length to be measured (resulting largely from the finite accuracy with which these features may be located in the image) is modelled by covariance matrices $\Lambda_{\mathbf{x}}$ and $\Lambda_{\mathbf{x}'}$.

Maximum likelihood estimation of the end points and uncertainties.

In this section the P matrix is assumed noise-free; this assumption is removed below (5.26).

Since in the error-free case, $\mathbf{x}$ and $\mathbf{x}'$ must be aligned with the vertical vanishing point $\mathbf{v}$, the maximum likelihood estimates ($\hat{\mathbf{x}}$ and $\hat{\mathbf{x}}'$) of their true

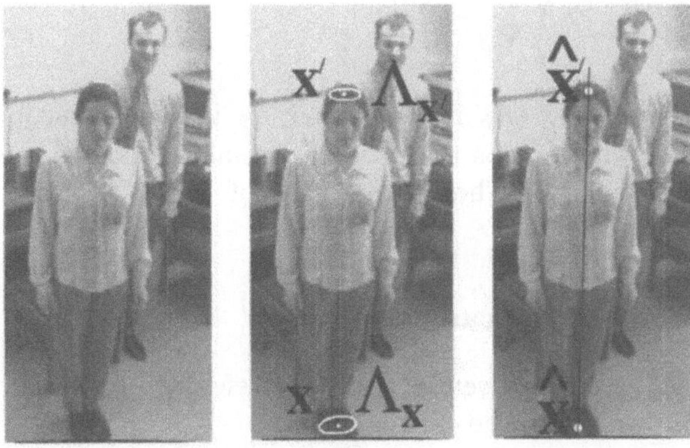

a b c

Fig. 5.14. Maximum likelihood estimation of the end points. **a** Original image (closeup of Fig. 5.15). **b** The uncertainty ellipses of the end points, $\Lambda_{\mathbf{x}}$ and $\Lambda_{\mathbf{x}'}$, are shown. These ellipses are defined manually; they indicate a 3-standard deviation confidence region for localizing the points. **c** The computed MLE end points $\hat{\mathbf{x}}$ and $\hat{\mathbf{x}}'$ are aligned with the vertical vanishing point (outside the image).

locations can be determined by minimizing the sum of the Mahalanobis distances between the input points $\mathbf{x}$ and $\mathbf{x}'$ and their MLE estimates $\hat{\mathbf{x}}$ and $\hat{\mathbf{x}}'$:

$$\min_{\hat{\mathbf{x}}_2, \hat{\mathbf{x}}'_2,} \left[(\mathbf{x}_2 - \hat{\mathbf{x}}_2)^\top \Lambda_{\mathbf{x}_2}^{-1} (\mathbf{x}_2 - \hat{\mathbf{x}}_2) + (\mathbf{x}'_2 - \hat{\mathbf{x}}'_2)^\top \Lambda_{\mathbf{x}'_2}^{-1} (\mathbf{x}'_2 - \hat{\mathbf{x}}'_2) \right] \tag{5.24}$$

subject to the *alignment constraint*

$$\mathbf{v} \cdot (\hat{\mathbf{x}} \times \hat{\mathbf{x}}') = 0 \tag{5.25}$$

(the subscript 2 indicates inhomogeneous 2-vectors).

This is a constrained minimization problem. A closed form solution can be found (by the Lagrange multiplier method) in the special case where

$$\Lambda_{\mathbf{x}'_2} = \gamma^2 \Lambda_{\mathbf{x}_2}$$

with γ a scalar, but unfortunately in the general case the problem has no closed-form solution. Nevertheless, in the general case, an initial solution can be computed by using the approximation given in Appendix B and then refining it by running a numerical algorithm, such as the Levenberg-Marquardt.

Once the MLE end points have been estimated, standard techniques [20, 38] are employed to obtain a first order approximation to the 4 $\times$ 4, rank-three covariance of the MLE 4-vector $\hat{\boldsymbol{\zeta}}^\top = (\hat{\mathbf{x}}_2'^\top \; \hat{\mathbf{x}}_2^\top)$. Figure 5.14 illustrates the idea (see appendix C for details).

Uncertainty of distance measurements.
Assuming noise in both the end points and the projection matrix, and statistical independence between $\hat{\boldsymbol{\zeta}}$ and $\mathtt{P}$ we obtain a first order approximation for the variance of the distance Z of a point from a plane:

$$\sigma_Z^2 = \nabla_Z \begin{pmatrix} \Lambda_{\hat{\zeta}} & 0 \\ 0 & \Lambda_P \end{pmatrix} \nabla_Z^\top \tag{5.26}$$

where ∇_Z is the 1×10 ($\Lambda_{\hat{\zeta}}$ is 4×4 and Λ_P is 6×6) Jacobian matrix of the function (5.9) which maps the projection matrix and the end points $\mathbf{x}$, $\mathbf{x}'$ to their world distance Z. The computation of ∇_Z is explained in detail in Appendix C.

5.3.3 Uncertainty of the camera position

The distance of the camera centre from the reference plane is computed according to (5.20) which can be rewritten as:

$$Z_c = -(\mathbf{p}_4 \cdot \mathbf{p}_3)^{-1} \tag{5.27}$$

If an exact P matrix is assumed, then the camera distance is exact too, in fact it depends only on the matrix elements of P. Similarly, the accuracy of Z_c depends only on the accuracy of the P matrix.

Equation (5.27) maps $\mathcal{R}^6$ into $\mathcal{R}$, and the associated 1×6 Jacobian matrix ∇Z_c is readily derived to be:

$$\nabla Z_c = Z_c^2 \left(\mathbf{p}_4^\top \; \mathbf{p}_3^\top \right)$$

and, from a first order analysis the variance of Z_c is:

$$\sigma_{Z_c}^2 = \nabla Z_c \Lambda_P \nabla Z_c^\top \tag{5.28}$$

where Λ_P is computed in section 5.3.1.

The variances $\sigma_{X_c}^2$ and $\sigma_{Y_c}^2$ of the X, Y location of the camera can be computed in a similar way [24].

5.3.4 Example – uncertainty of heights of people

In this section we show the effects of the number of reference distances and image localization error on the predicted uncertainty in measurements.

An image obtained from a poor-quality security camera is shown in Fig. 5.15**a**. It has been corrected for radial distortion using the method described in section 3.2, and the floor taken as the reference plane. Vertical and horizontal lines are used to compute the P matrix of the scene. The vanishing line for the ground plane is shown in white at the top of each image.

The scene is calibrated by identifying two points $\mathbf{v}_1, \mathbf{v}_2$ on the reference plane's vanishing line (shown in white at the top of each image) and the vertical vanishing point $\mathbf{v}$. These points are computed by intersecting sets of parallel lines. The uncertainty of each point is assumed to be Gaussian and isotropic with standard deviation 0.1 pixels. The uncertainty of the vanishing line is derived from a first order propagation through the vector product operation $\mathbf{l} = \mathbf{v}_1 \times \mathbf{v}_2$. The projection matrix P is therefore uncertain with its covariance given by (5.21).

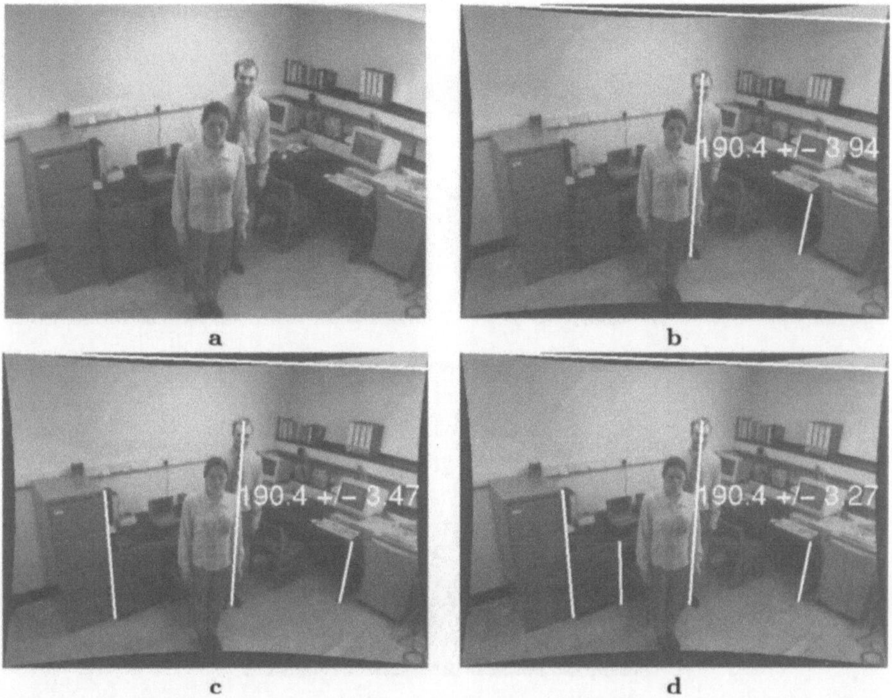

Fig. 5.15. Measuring heights and estimating their uncertainty. **a** Original image. **b** Image corrected for radial distortion and measurements superimposed. With only *one* supplied reference height the man's height has been measured to be $Z = 190.4 \pm 3.94$ cm, (cf. ground truth value, 190 cm). The uncertainty has been estimated by using (5.26) (the uncertainty bound is at $\pm$ 3 standard deviations). **c** With *two* reference heights $Z = 190.4 \pm 3.47$ cm. **d** With *three* reference heights $Z = 190.4 \pm 3.27$ cm. In the limit $\Lambda_{\mathtt{P}} = 0$ (error-free $\mathtt{P}$ matrix), the height uncertainty reduces to 2.16 cm for all (**b, c, d**); the residual error, in this case, is due only to the error on the two end points.

In addition, the end points of the height to be measured are assumed to be uncertain and their covariances estimated as in section 5.3.2. The uncertainties in the height measurements shown are computed as 3-standard deviation intervals.

In Fig. 5.15**b** one reference height is used to compute the affine scale factor α from (5.8) (the minimum number of references is used). Uncertainty has been assumed in the reference heights, vertical vanishing point and plane vanishing line. Once α is computed, other measurements in the same direction are metric. The height of the man has been computed and shown in the figure. It differs by 4 mm from the known true value.

The uncertainty associated with the height of the man is computed from (5.26) and displayed in Fig. 5.15**b**. Note that the true height value falls always within the computed 3-standard deviation range, as expected.

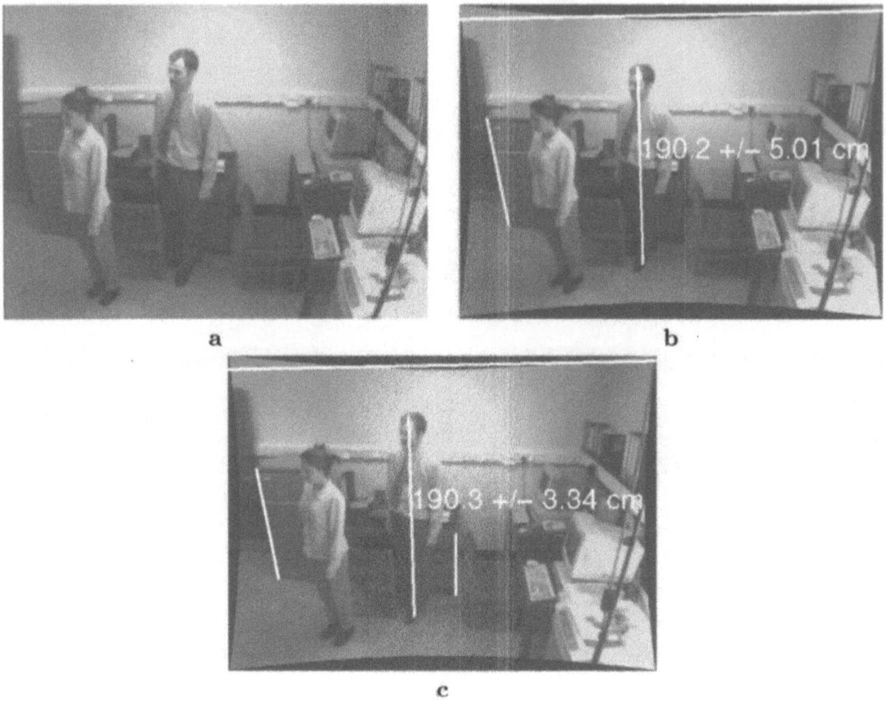

Fig. 5.16. Measuring heights and estimating their uncertainty, second point of view. **a** Original image. **b** The image has been corrected for radial distortion and height measurements computed and superimposed. With *one* supplied reference height Z = 190.2 ± 5.01 cm (cf. ground truth value 190 cm). **c** With *two* reference heights Z = 190.4 ± 3.34 cm. Compare with Fig. 5.15.

As the number of reference distances is increased (see Figs 5.15**c** and **d**), the uncertainty of P (in fact, just on α) decreases, resulting in a decrease in uncertainty of the measured height, as theoretically expected (see Appendix C for details). Equation (5.11) has been employed here to metric calibrate the distance from the floor.

Figure 5.16 shows images of the same scene with the same people, but acquired from a different point of view. As before, the uncertainty of the measurements decreases as the number of references increases (Fig. 5.16**b** and **c**). The measurement is the same as in the previous view (Fig. 5.15) thus demonstrating invariance to camera location.

Figure 5.17 shows an example where the height of the woman and the related uncertainty are computed for two different orientations of the uncertainty ellipses of the end points. In Fig. 5.17**b** the two input ellipses of Fig. 5.17**a** have been rotated by an angle of approximately 40°, maintaining the size and position of the centres. The angle between the direction defined by the major axes (direction of maximum uncertainty) of each ellipse and the measuring direction

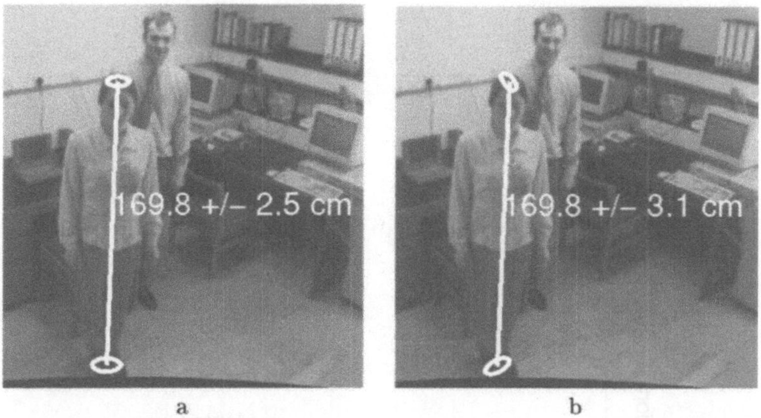

Fig. 5.17. Estimating the uncertainty in height measurements for different orientations of the input 3-standard deviation uncertainty ellipses. **(a)** Cropped version of image 5.15**b** with measurements superimposed: $Z = 169.8 \pm 2.5$ cm (at three standard deviations). The ground truth is $Z = 170$ cm, it lies within the computed range; **(b)** the input ellipses have been rotated keeping their size and position fixed: $Z = 169.8 \pm 3.1$ cm (at three standard deviations). The height measurement is less accurate.

is smaller than in Fig. 5.17**a** and the uncertainty in the measurements greater, as expected.

5.3.5 Validation of uncertainty analysis

In this section the first order error analysis previously described is validated by computing the uncertainty of the height of the man in Fig. 5.15**e** using Monte Carlo simulations as described in Table 5.1 and comparing it to the first order analytical result. Specifically, the statistical standard deviation of the man's height from a reference plane is computed and compared with the standard deviation obtained from the first order error analysis.

Uncertainty is modelled as Gaussian noise and described by covariance matrices (see section 3.4). Noise is assumed on the end points of the three reference distances and also on the vertical vanishing point, the plane vanishing line and on the end points of the height to be measured.

Figure 5.18 shows the results of the test. The base point is randomly distributed according to a 2D non-isotropic Gaussian about the mean location $\mathbf{x}$ (on the feet of the man in Fig. 5.15) with covariance matrix $\Lambda_{\mathbf{x}}$ (Fig. 5.18**a**). Similarly the top point is randomly distributed according to a 2D non-isotropic Gaussian about the mean location $\mathbf{x}'$ (on the head of the man in Fig. 5.15), with covariance $\Lambda_{\mathbf{x}'}$ (Fig. 5.18**b**).

The two covariance matrices are respectively:

$$\Lambda_{\mathbf{x}} = \begin{pmatrix} 10.18 & 0.59 \\ 0.59 & 6.52 \end{pmatrix} \qquad \Lambda_{\mathbf{x}'} = \begin{pmatrix} 4.01 & 0.22 \\ 0.22 & 1.36 \end{pmatrix}$$

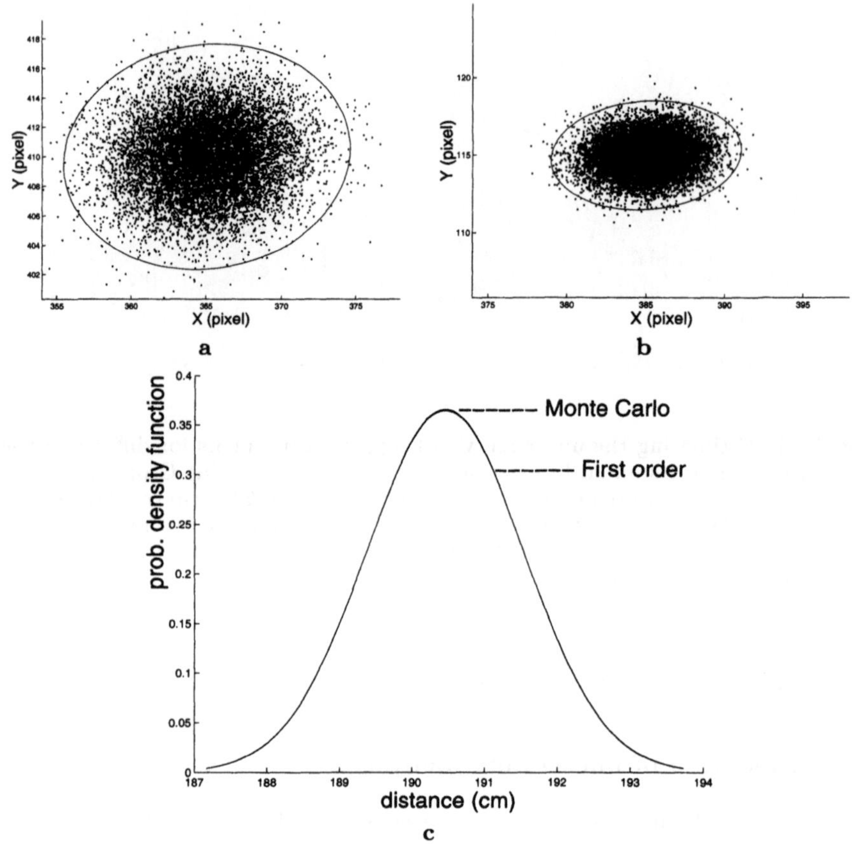

Fig. 5.18. Monte Carlo simulation of the example in Fig. 5.15**d**. **a** Distribution of the input base point $\mathbf{x}$ and the corresponding 3-standard deviation ellipse. **b** Distribution of the input top point $\mathbf{x}'$ and the corresponding 3-standard deviation ellipse. Figures **a** and **b** are drawn at the same scale. **c** The analytical and simulated distributions of the computed distance Z. The two curves are almost identical.

Suitable values for the covariances of the three references, the vanishing point and the vanishing line have been used. The simulation has been run with $N = 10\,000$ samples.

Analytical and simulated distributions of Z are plotted in Fig. 5.18**c**; the two curves are almost identical. Any slight difference is due to the assumptions of statistical independence (5.21, 5.22 and 5.26) and first order truncation introduced by the error analysis.

A comparison between the statistical and the analytical standard deviations is reported in the table below with the corresponding relative error:

First order	Monte Carlo	Relative error
σ_Z	σ'_Z	$\frac{\lvert\sigma_Z - \sigma'_Z\rvert}{\sigma'_Z}$
1.091 cm	1.087 cm	0.37%

- For j=1 to N (with N = number of samples)
  - For each reference: given the measured reference end points $\mathbf{r}$ (on the reference plane) and $\mathbf{r}'$, generate a random base point $\mathbf{r}_j$, a random top point $\mathbf{r}'_j$ and a random reference distance $\mathbf{Z}_{r_j}$ according to the associated covariances.
  - Generate a random vanishing point according to its covariance $\Lambda_{\mathbf{v}}$.
  - Generate a random plane vanishing line according to its covariance $\Lambda_{\mathbf{l}}$.
  - Compute the α parameter by applying (5.11) to the references, and the current P matrix (5.3).
  - Generate a random base point $\mathbf{x}_j$ and a random top point $\mathbf{x}'_j$ for the distance to be computed according to their respective covariances $\Lambda_{\mathbf{x}}$ and $\Lambda_{\mathbf{x}'}$.
  - Project the points $\mathbf{x}_j$ and $\mathbf{x}'_j$ onto the best fitting line through the vanishing point (see section 5.3.2).
  - Compute the current distance Z_j by applying (5.9).
- The statistical standard deviation of the population of simulated Z_j values is computed as:

$$\sigma'^2_Z = \frac{\sum_{j=1}^{N}(Z_j - \overline{Z})^2}{N-1}$$

and compared to the analytical one (5.26).

Table 5.1. Monte Carlo simulation.

Note that the height $Z = 190.45$ cm and the associated first order uncertainty $3\sigma_Z = 3.27$ cm is shown in Fig. 5.15d.

In the limit $\Lambda_{\mathtt{P}} = 0$ (error-free P matrix), the simulated and analytical results are even closer.

This result shows the validity of the first order approximation in this case, and numerous other examples have followed the same pattern. However, some care must be exercised since as the input uncertainty increases, not only does the output uncertainty increase, but the relative error between the statistical and the analytical output standard deviations also increases. For large covariances, the assumption of linearity is poorly founded; therefore the first order analysis no longer holds.

This is illustrated in the table below where the relative error is shown for various increasing values of the input uncertainties. The uncertainties of the references distances and the end points are multiplied by the increasing factor γ. For instance, if $\Lambda_{\mathbf{x}}$ is the covariance of the image point $\mathbf{x}$ then $\Lambda_{\mathbf{x}}(\gamma) = \gamma^2\Lambda_{\mathbf{x}}$.

γ	1	5	10	20	30
$\frac{\lvert\sigma_Z - \sigma'_Z\rvert}{\sigma'_Z}$(%)	0.37	1.68	3.15	8.71	16.95

In the *affine case* (when the vertical vanishing point and the plane vanishing line are at infinity) the first order error propagation is exact (no longer just an approximation as in the general projective case), and the analytic and simulated results coincide.

5.4 Three-dimensional metrology from a single view

This section describes a very simple but effective generalization of the theory described above.

The single-view techniques for measuring distances between two planes in a pencil are extended to making full three-dimensional measurements in the observed scene. The key idea is a new description of the three-dimensional space as a collection of three pencils of parallel planes in three different directions [7]. That leads to a complete three-dimensional reconstruction from single images.

5.4.1 The three-dimensional space as a collection of pencils of parallel planes

If $\mathbf{v}_x$, $\mathbf{v}_y$ and $\mathbf{v}_z$ are the three vanishing points (with unit third coordinate) for three directions in space (not necessarily orthogonal to each other) and $\mathbf{o}$ is the origin of the image coordinate system, then the $\mathbf{P}$ matrix can be written as [38]:

$$\mathtt{P} = \left[\begin{array}{cccc} \alpha_x \mathbf{v}_x & \alpha_y \mathbf{v}_y & \alpha_z \mathbf{v}_z & \mathbf{o} \end{array}\right] \tag{5.29}$$

with α_x, α_y and α_z three unknown scale factors.

Equation (5.29) generalizes (5.3) and allows us to extend the theory discussed in section 5.2 to parametrizing the space as *three* pencils of planes rather than only one. Measurements can now be computed in three directions, thus specifying the complete three-dimensional affine position of world points.

As explained in section 5.2, attention must be paid to the fact that the image point $\mathbf{o}$ can be chosen arbitrarily but, in order to avoid degeneracies in the parametrization, it must not lie on any of the three vanishing lines identified by the three vanishing points $\mathbf{v}_x$, $\mathbf{v}_y$ and $\mathbf{v}_z$, i.e.

$$\mathbf{o} \cdot (\mathbf{v}_i \times \mathbf{v}_j) \neq 0 \quad \forall i, j = x, y, z$$

A possible choice of the point $\mathbf{o}$ may then be the average (centre of mass) of the three vanishing points in homogeneous coordinates:

$$\mathbf{o} = \frac{\sum_{i=x,y,z} \mathbf{v}_i}{3}$$

5.4.2 Three-dimensional location of points

Generalizing (5.8), the X, Y, Z location of a three-dimensional point $\mathbf{Q}$ may be computed from single images applying the following formulae (see Fig. 5.19):

$$\alpha_x X = -\frac{\mathbf{o} \cdot \mathbf{l}_{yz}}{\mathbf{q}_{yz} \cdot \mathbf{l}_{yz}} \frac{||\mathbf{q}_{yz} \times \mathbf{q}||}{||\mathbf{v}_x \times \mathbf{q}||} \tag{5.30}$$

$$\alpha_y Y = -\frac{\mathbf{o} \cdot \mathbf{l}_{xz}}{\mathbf{q}_{xz} \cdot \mathbf{l}_{xz}} \frac{||\mathbf{q}_{xz} \times \mathbf{q}||}{||\mathbf{v}_y \times \mathbf{q}||}$$

$$\alpha_z Z = -\frac{\mathbf{o} \cdot \mathbf{l}_{xy}}{\mathbf{q}_{xy} \cdot \mathbf{l}_{xy}} \frac{||\mathbf{q}_{xy} \times \mathbf{q}||}{||\mathbf{v}_z \times \mathbf{q}||}$$

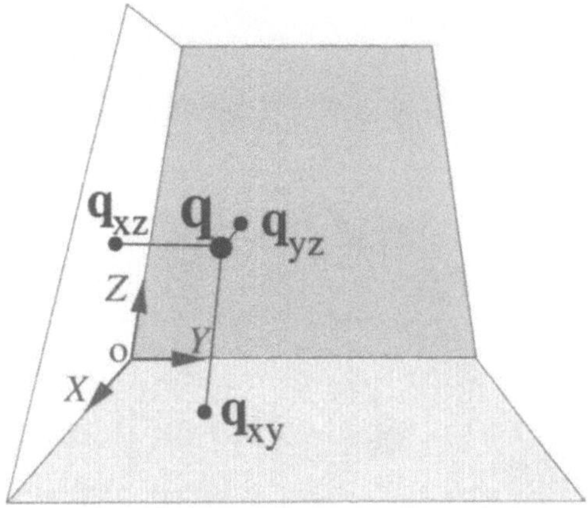

Fig. 5.19. Complete affine three-dimensional reconstruction from single images. The three-dimensional space is parametrized as three pencils of parallel planes (only one plane for each pencil is shown, the three pencils do not need to be orthogonal to each other). The location of a world point $\mathbf{Q}$ can be computed by applying (5.30) (see also (5.8)).

Where $\mathbf{q}$ is the image of the world point $\mathbf{Q}$, $\mathbf{l}_{ij} = \mathbf{v}_i \times \mathbf{v}_j$ and $\mathbf{q}_{ij}$ is the intersection of the plane spanned by the axis i and j with the line through $\mathbf{Q}$ parallel to the direction k ($\forall i, j, k = x, y, z$). If the α_i parameters are known, then metric structure can be computed for the point $\mathbf{Q}$; otherwise only affine structure can be obtained.

Only one of the points $\mathbf{q}_{xy}$, $\mathbf{q}_{yz}$ and $\mathbf{q}_{xz}$ is necessary.
In images of real scenes it is quite rare to find a situation in which the three base points $\mathbf{q}_{xy}$, $\mathbf{q}_{yz}$ and $\mathbf{q}_{xz}$ are all visible. We now show that only one of those points needs to be known and that the other two can be computed.

For instance, it is easy to prove that if $\mathbf{q}_{xy}$ is known and $\mathbf{a}_x$ and $\mathbf{a}_y$ are the images of the chosen world coordinates axes X and Y, respectively, then:

$$\mathbf{q}_{yz} = (\mathbf{q} \times \mathbf{v}_x) \times (\mathbf{v}_z \times (\mathbf{a}_y \times (\mathbf{q}_{xy} \times \mathbf{v}_x))) \tag{5.31}$$

$$\mathbf{q}_{xz} = (\mathbf{q} \times \mathbf{v}_y) \times (\mathbf{v}_z \times (\mathbf{a}_x \times (\mathbf{q}_{xy} \times \mathbf{v}_y))) \tag{5.32}$$

The image points $\mathbf{q}_{yz}$ and $\mathbf{q}_{xz}$ can also be constructed geometrically as illustrated in Fig. 5.20. These operations correspond, in the world, to parallel projections of the three-dimensional point $\mathbf{Q}$ into the plane YZ along the X-direction in (5.31); and into the plane XZ along the Y-direction in (5.32). Symmetrical formulae can be derived in cases where $\mathbf{q}_{yz}$ or $\mathbf{q}_{xz}$ are known.

Equations (5.31) and (5.32) (or the symmetrical ones) may be substituted in (5.30) to obtain formulae which depend on one base point only.

5.4.3 Three-dimensional location of the camera

Equations (5.18) to compute the camera location still hold. But they can be rewritten in the light of this new parametrization, in the case of unknown α parameters as:

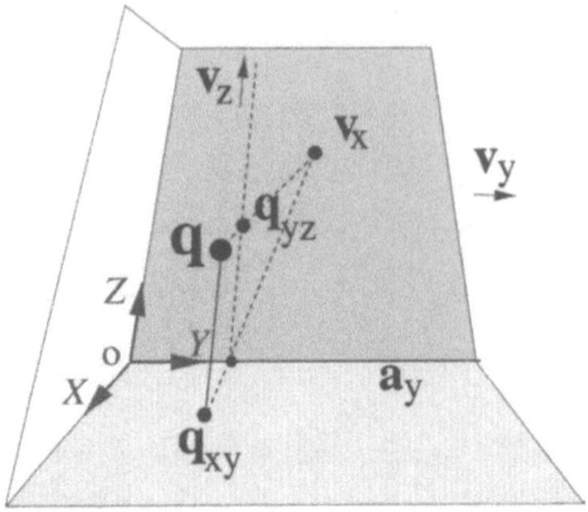

Fig. 5.20. Geometric construction of the point $\mathbf{q}_{yz}$: the points $\mathbf{q}$, $\mathbf{q}_{xy}$, $\mathbf{v}_x$, $\mathbf{v}_z$ and the line $\mathbf{a}_y$ are known and the point $\mathbf{q}_{yz}$ is constructed by line intersections. A similar construction applies for the point $\mathbf{q}_{xz}$.

$$\alpha_x X_c = -\det\left[\, \mathbf{v}_y \; \mathbf{v}_z \; \mathbf{p}_4 \,\right], \quad \alpha_y Y_c = \det\left[\, \mathbf{v}_x \; \mathbf{v}_z \; \mathbf{p}_4 \,\right],$$
$$\alpha_z Z_c = -\det\left[\, \mathbf{v}_x \; \mathbf{v}_y \; \mathbf{p}_4 \,\right], \quad W_c = \det\left[\, \mathbf{v}_x \; \mathbf{v}_y \; \mathbf{v}_z \,\right] \tag{5.33}$$

and thus the affine location of the camera centre is computed. As before, metric structure is obtained if the α parameters are known.

5.4.4 Camera calibration from a single view

Full calibration.
Given a world coordinate system XYZ with origin in $\mathbf{O}$ and an image coordinate system xy, if we know the three vanishing points for the X, Y and Z directions and at least one world distance measurement along each direction, by applying (5.30) the three calibration parameters α_i can be computed (see section 5.2.1). We can choose the origin $\mathbf{o}$ to be the image of $\mathbf{O}$ and from (5.29) the full world-to-image projection matrix $\mathtt{P}$ is determined and the scene calibrated. In fact the origin $\mathbf{o}$ and each vanishing point provide two degrees of freedom. The three scalars α_i provide three more degrees of freedom for a total of eleven independent degrees of freedom. Full external camera calibration is, therefore, obtained from a single image.

Once the scene has been calibrated, metric structure of points and cameras can be recovered from (5.30) and (5.33), respectively. Moreover, the three image-to-world homographies for each of the planes can be extracted directly from (5.29).

If the homography between a world plane (e.g. the plane XY) and the image is known, then only the α parameter for the direction not on that plane (e.g. α_z) must be known to compute a complete metric three-dimensional reconstruction, since computing the two remaining calibration parameters (α_x and α_y) from the known homography is straightforward.

Partial calibration.
There may be situations where only some of the calibration parameters α are known. In that case we obtain an intermediate, *partially affine* and *partially metric* structure. Partial calibration leads to a partial reconstruction of the scene. For instance, the situation analyzed in section 5.2.2 is characterized by affine calibration on the plane and metric off the plane (if α is known). This is sufficient to compute metric distances off the reference plane (but only affine measurements on the plane).

The theory developed in section 5.2.1 concerning the use of homologies to transfer measurements and points from one plane to a parallel one, and in section 5.3 concerning the uncertainty analysis on distances between planes still applies. It can be employed to measurements within the same pencil of planes.

5.5 Applications

The validity of the metrology algorithm presented here is demonstrated in this section with a number of practical applications.

5.5.1 Forensic science

A common requirement in surveillance images is to obtain measurements from the scene, such as the height of a suspect. Although the suspect has usually left the scene, reference lengths can be measured from fixtures such as tables and windows.

Figure 5.21 illustrates the computation of the height of the person standing next to a telephone box. The ground is the reference plane and the vertical is the reference direction. The edges of the paving stones are used to compute the vanishing line of the plane, the edges of the telephone box to compute the vertical vanishing point, and its height provides the metric calibration in the vertical direction (Fig. 5.21**b**). Only one reference height (minimal set) has been used in this example. The height of the person in the photograph is then computed using (5.9) and shown in Fig. 5.21**c**. The ground truth is 187 cm, but note that the person is leaning slightly down on his right foot.

The associated uncertainty has also been estimated; two uncertainty ellipses have been defined manually, one on the head of the person and the other on the feet and then propagated across the chain of computations as described in section 5.3 to give the 2.2 cm three-standard deviation uncertainty range shown in Fig. 5.21**c**.

5.5.2 Furniture measurements

In this section another application is described. Heights of furniture such as shelves, tables or windows in an indoor environment are measured.

Fig. 5.21. The height of a person standing by a public telephone is computed. **a** Original image. **b** The ground plane is the reference plane, and its vanishing line is computed from the paving stones on the floor. The vertical vanishing point is computed from the edges of the telephone box, whose height is known, and used as reference. The vanishing line and the reference height are shown. **c** The computed height of the person and the estimated uncertainty are shown. The veridical height is 187 cm, but note that the person is leaning slightly on his right foot.

Figure 5.22**a** shows a desk in The Queen's College upper library in Oxford. The floor is the reference plane and its vanishing line has been computed from intersecting edges of the floorboards. The vertical vanishing point has been computed by intersecting the vertical edges of the bookshelf. The vanishing line is shown in Fig. 5.22**b** with the reference height used.

The computed heights and associated uncertainties are shown in Fig. 5.22**c**. The uncertainty bound is ± 3 standard deviations. Note that the ground truth always falls within the computed uncertainty range. The height of the camera is computed as 1.71 m from the floor.

5.5.3 Virtual modelling

Figure 5.23 shows an example of a complete three-dimensional reconstruction of a real scene from a single image. Two sets of horizontal edges are used to compute the vanishing line for the ground plane, and vertical edges are used to compute the vertical vanishing point.

The height of the window frame and the height of one of the pillars are used as reference heights. Furthermore, the two sides of the base of the porch have been measured, thus defining the metric calibration of the ground plane.

Figure 5.23**b** shows a view of the reconstructed model. The person is represented as a flat silhouette. The position of the camera centre is also estimated and superimposed on a different view of the three-dimensional model in Fig. 5.23**c**.

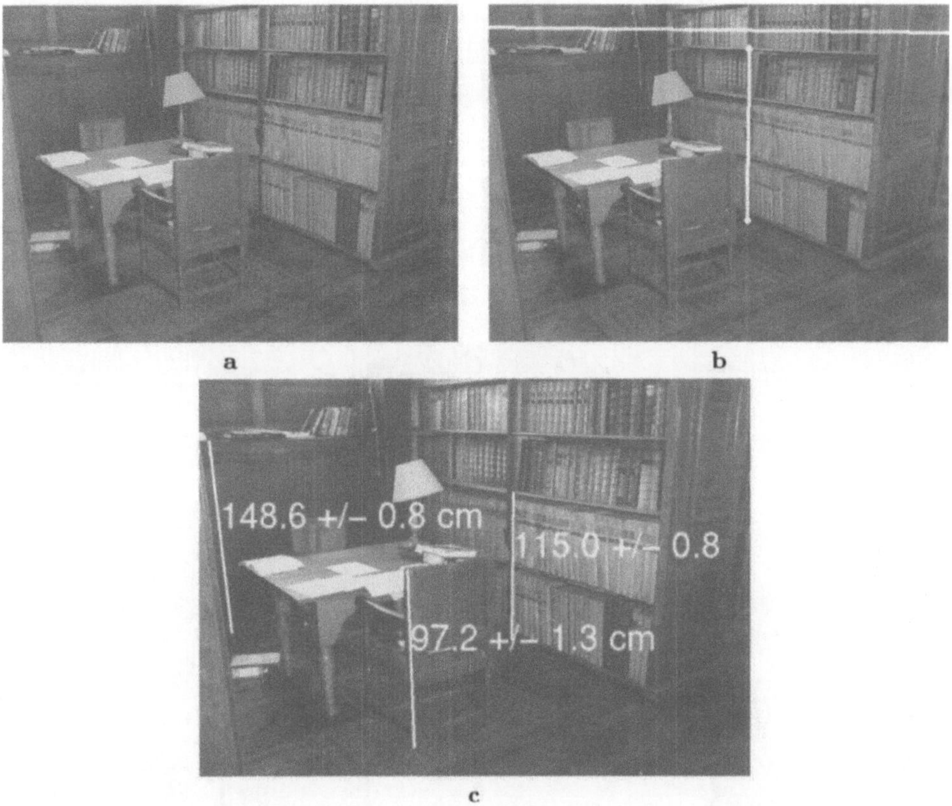

Fig. 5.22. Measuring the height of furniture in The Queen's College upper library, Oxford. **a** Original image. **b** The plane vanishing line (white horizontal line) and reference height (white vertical line) are superimposed on the original image; the marked shelf is 156 cm high. **(c)** Computed heights and related uncertainties; the uncertainty bound is at ±3 standard deviations. The ground truth is: 115 cm for the right hand shelf, 97 cm for the chair and 149 cm for the shelf on the left. Note that the ground truth always falls within the computed uncertainty range.

5.5.4 Modelling paintings

Figure 5.24 shows a masterpiece of Italian Renaissance painting, *Flagellazione* by Piero della Francesca (1416–1492). The painting, thanks to the artist's famous mathematical skills [32], faithfully follows the geometric rules of linear perspective, and therefore the techniques described can be applied to compute the structure of the scene.

Unlike other techniques [59], whose main aim is to create convincing new views of the painting regardless of the correctness of the three-dimensional geometry, here a geometrically correct model of the viewed scene is reconstructed (see Fig. 5.24**c** and **d**).

In the painting analyzed here, the ground plane is chosen as the reference and its vanishing line computed from the several parallel lines on it. The verti-

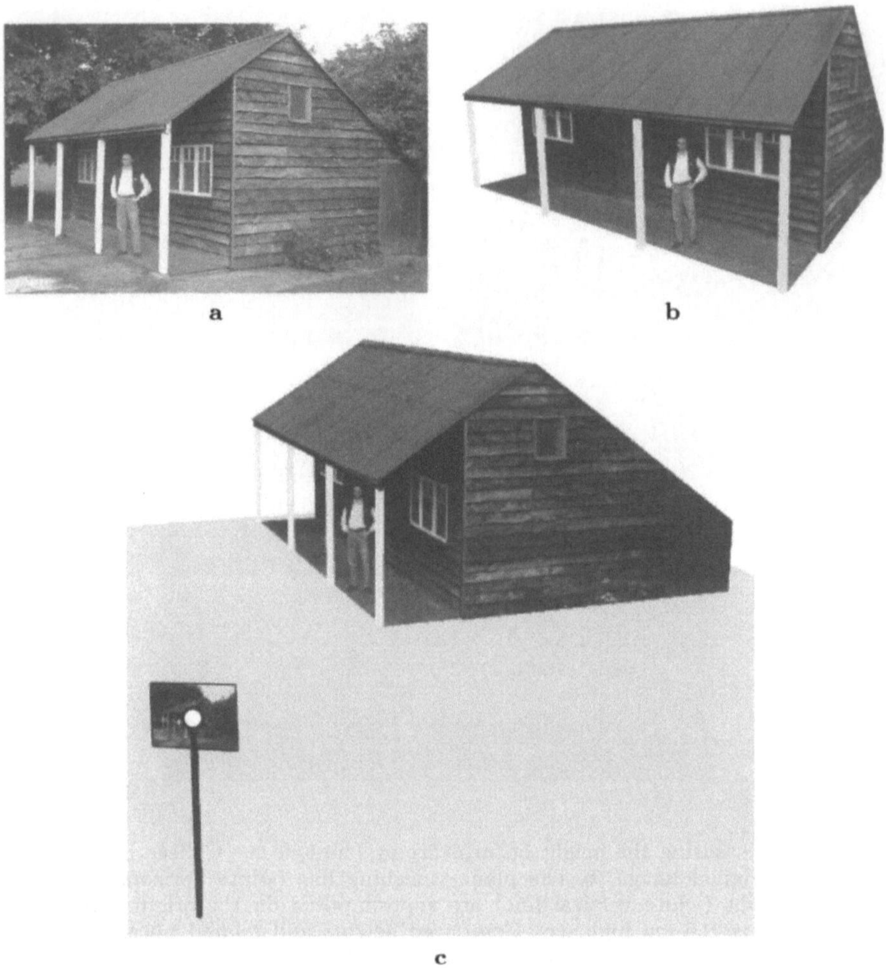

Fig. 5.23. A complete three-dimensional reconstruction of a scene from a photograph: **a** original image; **b** a view of the reconstructed three-dimensional model; **c** a view of the reconstructed three-dimensional model which shows the camera position and height with respect to the scene. These results are also shown in colour plate 1.

cal vanishing point follows from the vertical lines and consequently the relative heights of people and columns are defined. Figure 5.24**b** shows the painting with height measurements superimposed. Christ's height is taken as the reference and the heights of the other people are expressed as relative percentage differences. Notice the consistency between the height of the people in the foreground with the height of the people in the background.

By assuming a square floor pattern, the ground plane is rectified and the position of each object estimated [24, 77]. The scale of the floor relative to heights is set from the ratio between the height and the base of the frontoparallel

Fig. 5.24. A complete three-dimensional reconstruction of a Renaissance painting. **a** The painting *Flagellazione* (approx. 1453), by Piero della Francesca (1416–1492), Galleria Nazionale delle Marche, Urbino, Italia. Courtesy of Ministero per i Beni e le Attività Culturali. **b** Height measurements are superimposed on the original image. Christ's height is taken as the reference and the heights of all the other people are expressed as percent differences. The vanishing line is dashed. **c** A view of the reconstructed model. The patterned floor has been reconstructed in areas where it is occluded by taking advantage of the symmetry of its pattern. **d** Another view of the model with the roof removed to show the relative positions of people and architectural elements in the scene. Notice the repeated geometric pattern on the floor in the area delimited by the columns (barely visible in the painting). The people are represented simply as flat silhouettes since it is not possible to recover their volume from one image; they have been cut out manually from the original image. The columns have been approximated with cylinders. These results are also shown in colour plate 4 and colour plate 5.

archway. The measurements, up to an overall scale factor, are used to compute a three-dimensional model of the scene (scaled Euclidean structure).

Figure 5.24**c** shows a view of the reconstructed model. The people are represented again as flat silhouettes. The columns have been approximated with cylinders. The partially visible ceiling has been reconstructed correctly. Figure 5.24**d** shows a different view of the reconstructed model, where the roof has been removed to show the relative position of the people in the scene.

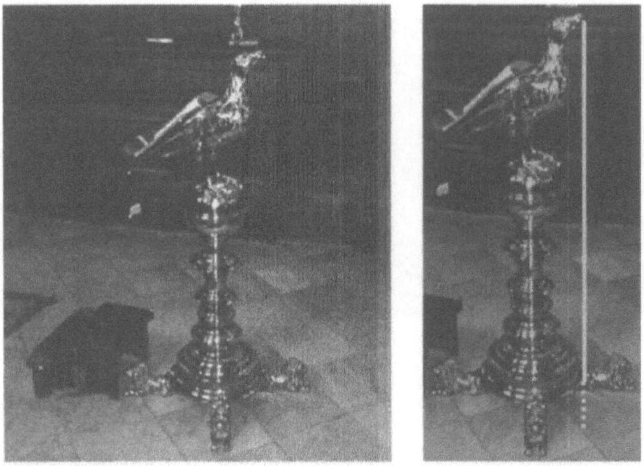

a b

Fig. 5.25. Single-view metrology cannot be applied when the base point is not defined: **a** original image – the lectern in The Queen's College chapel; **b** we want to measure the distance of the beak of the eagle from the floor, but the intersection of the ground plane with the vertical through the tip of the beak is not defined in the image and therefore single-view metrology cannot be applied.

5.6 Missing base point

A restriction on the single-view measurement algorithm presented in this chapter is the need to identify corresponding points between planes.

The problem.
One case where the method does not apply, therefore, is that of measuring the distance of a general three-dimensional point to a reference plane where the corresponding point on the floor in undefined (see Fig. 5.25). Here the homology is under-determined, and the line joining the image of this point to the vertical reference vanishing direction gives the locus of points which may correspond (therefore an uncertainty range for the height may still be computed).

Possible solutions.
A second view provides another locus, and the intersection of these loci in either view (transferred by the inter-image homography induced by the reference plane) uniquely defines the corresponding point on the reference plane. This construction was used in [100] to determine the vertical projection of a football onto the ground plane (see also [103] for a similar construction). The distance may then be computed as described in the main body of this chapter, or by using more direct formulae developed in the next chapter (see also [62]).

Here one intermediate case of interest is explored: when only one view is provided and a light-source casts shadows onto the reference plane. The light-source provides restrictions analogous to a second viewpoint [128], thus the projection (in the reference direction) of the three-dimensional point onto the reference plane may be determined as the intersection of a pair of one-dimensional loci (Fig. 5.26). In [101] this method was used to compute the trajectory of a football.

5.6.1 Using shadows

We wish to compute the distance of a three-dimensional point $\mathbf{X}$ from a reference plane but the intersection of the vertical line through $\mathbf{X}$ with the plane (base point $\mathbf{X}'$) is not defined (Fig. 5.26**a** shows a schematic of the situation described as it appears on the image plane). The single-view metrology algorithm is no use here, unless the image base point $\mathbf{x}'$ can be computed.

A reference height needs to be considered where top and base points can both be identified (points $\mathbf{r}$ and $\mathbf{r}'$ in Fig. 5.26**a**). The light source $\mathbf{s}$ casts a shadow $\mathbf{x}_s$ on the reference plane for the point $\mathbf{x}$ and similarly it casts a shadow $\mathbf{r}_s$ for the reference point $\mathbf{r}$. The point $\mathbf{r}'$ corresponds to $\mathbf{r}$ along the reference direction.

By representing points as homogeneous 3-vectors, the intersection of the line $< \mathbf{XR} >$ in space with the reference plane can be computed in the image as $\mathbf{p} = (\mathbf{x} \times \mathbf{r}) \times (\mathbf{x}_s \times \mathbf{r}_s)$ (referred to as the *piercing point*, see Fig. 5.26**b**). If we assume the light source is at infinity, then the two shadow lines $< \mathbf{R}'\mathbf{R}_s >$ and $< \mathbf{X}'\mathbf{X}_s >$ are parallel to each other in space and the point $\mathbf{v}_h$ is the vanishing point of their direction in the image plane. The point $\mathbf{v}_h$ can be computed by intersecting the reference shadow line and the vanishing line: $\mathbf{v}_h = (\mathbf{r}_s \times \mathbf{r}') \times \mathbf{l}$. Therefore the image point $\mathbf{x}'$ on the plane can be computed as $\mathbf{x}' = (\mathbf{x}_s \times \mathbf{v}_h) \times (\mathbf{p} \times \mathbf{r}')$. The algorithm described in section 5.2.1 may now be applied to compute the world distance between $\mathbf{x}$ and $\mathbf{x}'$. Furthermore, the vanishing point $\mathbf{v}$ for the reference direction may easily be computed as $\mathbf{v} = (\mathbf{r} \times \mathbf{r}') \times (\mathbf{x} \times \mathbf{x}')$.

The configuration described is analogous to the case of having one perspective and one affine camera described by Zhang et al. in [136].

Planar homology.
It is interesting to notice that the two triangles $\mathbf{r}, \mathbf{r}', \mathbf{r}_s$ and $\mathbf{x}, \mathbf{x}', \mathbf{x}_s$ are in a Desargues configuration and therefore they are related by a planar homology whose vertex is the piercing point and whose axis is the line $< \mathbf{s}\mathbf{v}_h >$ (Fig. 5.26**c**). Therefore, an alternative, algebraic way to compute the point $\mathbf{x}'$ is:

- estimate the homology $\mathtt{H}$ from its vertex (the point $\mathbf{p}$), its axis (the line $< \mathbf{s}\mathbf{v}_h >$) and a pair of corresponding points (e.g. $\mathbf{r}, \mathbf{x}$ or $\mathbf{r}_s, \mathbf{x}_s$) by applying (3.4);
- apply the homology projection to the reference point $\mathbf{r}'$:

$$\mathbf{x}' = \mathtt{H}\mathbf{r}' \tag{5.34}$$

From Desargues' theorem, the vanishing point $\mathbf{v}$ for the reference direction also lies on the axis of the homology: $\mathbf{v} \cdot (\mathbf{s} \times \mathbf{v}_h) = 0$. Often both $\mathbf{v}$ and $\mathbf{s}$ are known. In this case the position of the point $\mathbf{x}'$ is over-determined and it can be computed employing a maximum likelihood algorithm.

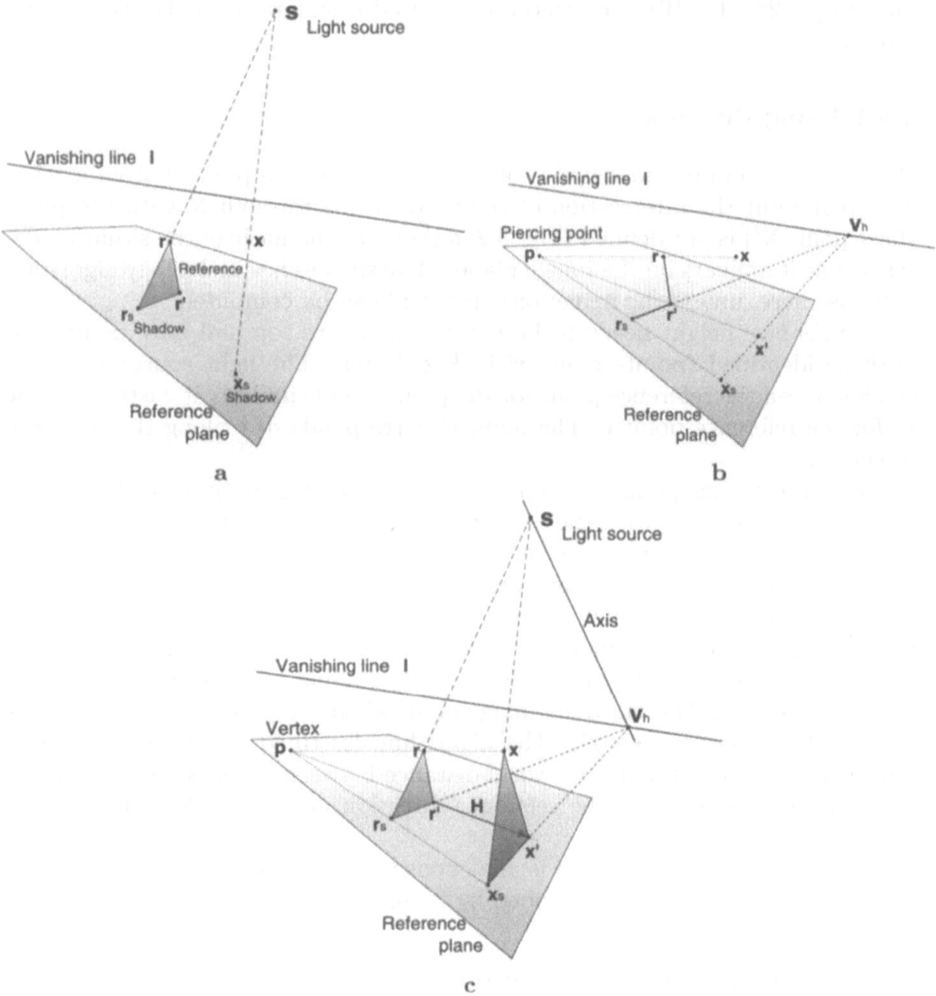

Fig. 5.26. Using shadows to compute distances in single views. **a** We wish to compute the distance of the point $\mathbf{x}$ from the reference plane but the base point ($\mathbf{x}'$) is not defined. The reference point $\mathbf{r}$ (off the plane) and its base $\mathbf{r}'$ (on the plane) correspond along the reference direction. The light source $\mathbf{s}$ (at infinity) casts the shadow points $\mathbf{r}_s$ and $\mathbf{x}_s$ onto the reference plane. **b** The position of the base point $\mathbf{x}'$ is computed by making use of the shadows on the reference plane (see text). **c** The two triangles $\mathbf{r}, \mathbf{r}', \mathbf{r}_s$ and $\mathbf{x}, \mathbf{x}', \mathbf{x}_s$ are in a Desargues configuration and therefore related by a planar homology $\mathbf{H}$. The piercing point $\mathbf{p}$ (image of the intersection of the line $< \mathbf{XR} >$ with the reference plane) is its vertex and the line joining the points $\mathbf{s}$ and $\mathbf{v}_h$ (on the plane at infinity) its axis.

The case of finite light source.
In the case where the light source is finite, then the two shadow lines $< \mathbf{R}'\mathbf{R}_s >$ and $< \mathbf{X}'\mathbf{X}_s >$ are no longer parallel to each other and therefore the vanishing point $\mathbf{v}_h$ cannot be computed.

Nevertheless if the vertical vanishing point $\mathbf{v}$ for the reference direction is known, then the axis of the homology is still defined: $\mathbf{a} = \mathbf{v} \times \mathbf{s}$; and the homology $\mathtt{H}$ can be computed. Hence the point $\mathbf{x}'$ can be computed from (5.34). Computing the world distance of $\mathbf{x}'$ from the reference plane is now straightforward.

6. Metrology from planar parallax

This chapter explores the geometry of two and three views in a *plane-plus-parallax* framework. Algorithms for structure and camera computation are developed and the results compared with the single-view metrology described in the previous chapter (see also [23]; similar, independent work can be found in [62]).

In particular, we address the problem of computing distances of points from planes when the intersection of the plane with the line through the point parallel to the reference direction is not defined (see Fig. 5.25). In such a case the single-view metrology approach is not sufficient. In section 5.6 we have shown how to overcome the problem when a light source casts a shadow onto a plane. In the case where no shadows can be detected then one more view provides the solution.

The plane-plus-points configuration has received significant attention in the past, not least because it arises frequently in everyday scenes. A useful and popular approach to the problem decomposes the image motion into a planar homographic transfer plus a residual image parallax vector [61, 62, 72, 105]. This decomposition has the advantage that it partially factors out dependence on the relative rotation and internal parameters of the camera.

In recent work, Carlsson and Weinshall *et al.* [13, 14, 131, 133] have demonstrated the fundamental duality of the three-dimensional reconstruction problem. They show that for points and cameras in *general position*, the problem of computing camera locations from n points in m views is mathematically equivalent to the problem of reconstructing $m + 4$ points in $n - 4$ views. This chapter analyzes, in particular, the geometry of *two* views of *seven* points, four of which are coplanar, and the geometry of *three* views of *six* points, four of which are coplanar. We also prove that the two configurations are dual, and that the fundamental geometric constraints in each case are encapsulated by a planar homology [106, 127]. A summary of the duality results contrasted with the general-position cases is shown in table 6.1.

The work here unifies a number of previously diverse results related to planar parallax [61, 72, 105], duality [13, 14, 131, 133] and planar homologies [127].

Formulae for computing the distance of the cameras from a distinguished world plane and formulae for structure computations, and associated uncertainty, are presented in the second part of this chapter, where the trifocal tensor [56, 111, 116] is also derived in the plane-plus-points case, showing that

m views	n pts	general position	coplanar
2	7	$3n+7=28$ dof $2mn=28$ constr. F determined up to a three-fold ambiguity No further constraints	$3n-1+7=27$ dof $2mn=28$ constraints F determined uniquely Motion constraint (one) in addition ∃ *homology*: maps between views, *vertex* is epipole, i.e. intersection of plane and camera baseline, *axis* is intersection of plane with plane containing remaining three points
3	6	$3n+18=36$ dof $2mn=36$ constr. T determined up to a three-fold ambiguity No further constraints	$3n-1+18=35$ dof $2mn=36$ constraints T determined uniquely Structure constraint (one) in addition ∃ *homology*: maps between points, *vertex* is intersection of plane and line joining the remaining two points, *axis* is intersection of plane and plane containing the camera centres.

Table 6.1. Camera/point duality results for: (i) points in general position and (ii) four points lying on a distinguished plane. The fundamental matrix F has seven degrees of freedom and the trifocal tensor T has eighteen degrees of freedom.

it is obtained uniquely. Structure and camera computations are obtained directly from image measurements, i.e. the parallax vectors, *without* needing to compute the epipolar geometry.

The remainder of the chapter is organized as follows. We begin with a discussion of background material, notation and parallax geometry. Then the geometry of two views, seven points, four of which are coplanar, is described. We show that there exists a homology on the plane relating the two views and derive the necessary conditions for the homology directly in terms of the parallax measurements. Section 6.2.2 shows the duality of the geometry of three views, six points to the two-view, seven-point case, and hence obtain the analogous necessary conditions. We also derive the trifocal tensor and show that it is over-constrained. Section 6.3 derives expressions for the distance of the cameras from the specified, reference plane, and the structure of points in terms of affine invariants; we also show how plane-plus-parallax geometry can be employed to compute the line of intersection between two planes. An uncertainty analysis on those measurements is performed in section 6.4 to estimate the confidence interval around each distance measurement. This analysis is validated, once again, by comparing the results with Monte Carlo statistical tests. Several examples on real images are presented.

6.1 Background

6.1.1 Notation

The area of a triangle with general vertices **a**, **b** and **c** is denoted $A_{\mathbf{abc}}$, and can be determined via the formula $A_{\mathbf{abc}} = \frac{1}{2}|\mathbf{abc}|$ where the points **a**, **b**, and **c** are represented as homogeneous 3-vectors with the last component equal to one.

Numbered subscripts are used to distinguish different views, with the first camera centre given by $\mathbf{O}_1$, the second by $\mathbf{O}_2$ and the third by $\mathbf{O}_3$. The projection of an image point onto the distinguished world plane from the ith view is denoted by $\mathbf{x}_i$ (see Fig. 6.1).

Superscripts identify projections of a three-dimensional point onto different views, thus the three-dimensional point $\mathbf{X}$ is imaged as $\mathbf{x}$ in the first image, $\mathbf{x}'$ in the second and $\mathbf{x}''$ in the third. The symbols $\dot{}$ and $\ddot{}$ indicate the homographic projection of a point from the second or third image into the first one, respectively; thus $\dot{\mathbf{x}} = \mathtt{H}_{21}\mathbf{x}'$ and $\ddot{\mathbf{x}} = \mathtt{H}_{31}\mathbf{x}''$, where $\mathtt{H}_{21}$ is the homography that maps points from the second image onto the first one and $\mathtt{H}_{31}$ is the homography that maps points from the third image onto the first one.

6.1.2 Planar parallax

The underlying parallax geometry is shown in Fig. 6.2; two views and a reference world plane are shown. The distinguished world plane induces a homography between the two views, meaning that the images of points on the plane can be transferred via the inter-image homography between views 1 and 2 (see section 3.1.2). The homography can be determined as usual from a minimum of four correspondences in the two views of points (or lines) on the distinguished plane.

The parallax vector in the first view is the vector joining the image $\mathbf{x}$ of a world point $\mathbf{X}$ with the transferred location $\dot{\mathbf{x}}$ of $\mathbf{X}$'s image in the other view (the point $\mathbf{x}'$). Furthermore, since the three planes (distinguished world plane and two image planes) are equivalent up to a plane projectivity we can also measure parallax in the second view or, if the image-to-world plane homographies are known, on the distinguished world plane. It is particularly advantageous to work with the world plane; in this case all dependence on the rotational and internal parameters of the cameras is removed (aggregated into the image-to-world plane homographies), leaving only a dependence on the location of the camera centres.

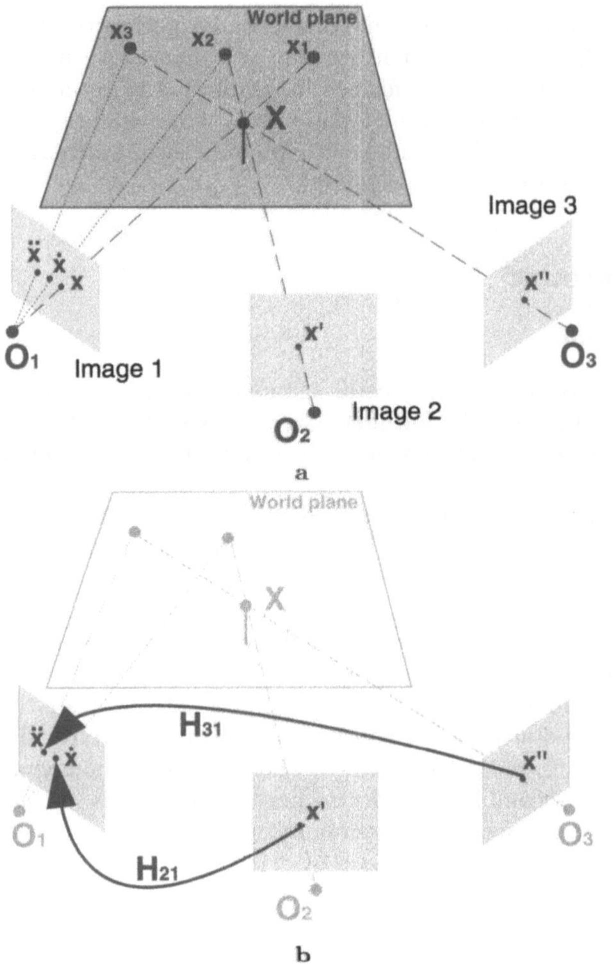

Fig. 6.1. The notation of points and camera centres. **a** Three cameras are viewing a slanted reference plane. The point $\mathbf{X}$ does not lie on the plane. Therefore, the projections of the point $\mathbf{X}$ on the plane from the three cameras are three distinct points: $\mathbf{x}_1$, $\mathbf{x}_2$ and $\mathbf{x}_3$. **b** Points $\dot{\mathbf{x}}$ and $\ddot{\mathbf{x}}$ on the first image are obtained by homographic projection from the second and third image respectively.

Since the clarity of the underlying geometry is greatly increased, we depict all relevant points and vectors on the world plane in all of the figures. However, the computations do *not* require the image-to-world homographies to be known and they can be carried out from images directly.

The parallax vector is directed towards (or away from) the epipole (see Fig. 6.2**b**), thus two such vectors are sufficient to compute its position, and the full epipolar geometry follows [6, 80, 85]. The magnitude of the parallax vector is related to the distance of the world point and cameras from the world

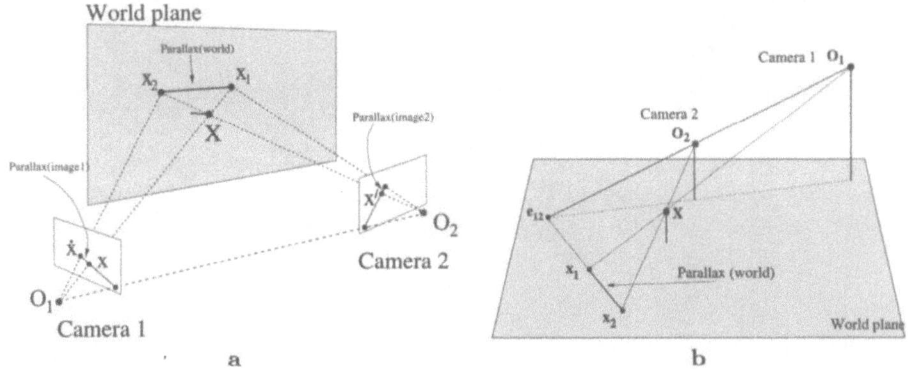

Fig. 6.2. Parallax geometry: **a** general configuration; parallax vectors are identified in both images and also on the world plane. **b** viewed only on the distinguished plane. The parallax vector $< \mathbf{x}_1\mathbf{x}_2 >$ passes through the epipole $\mathbf{e}_{12}$.

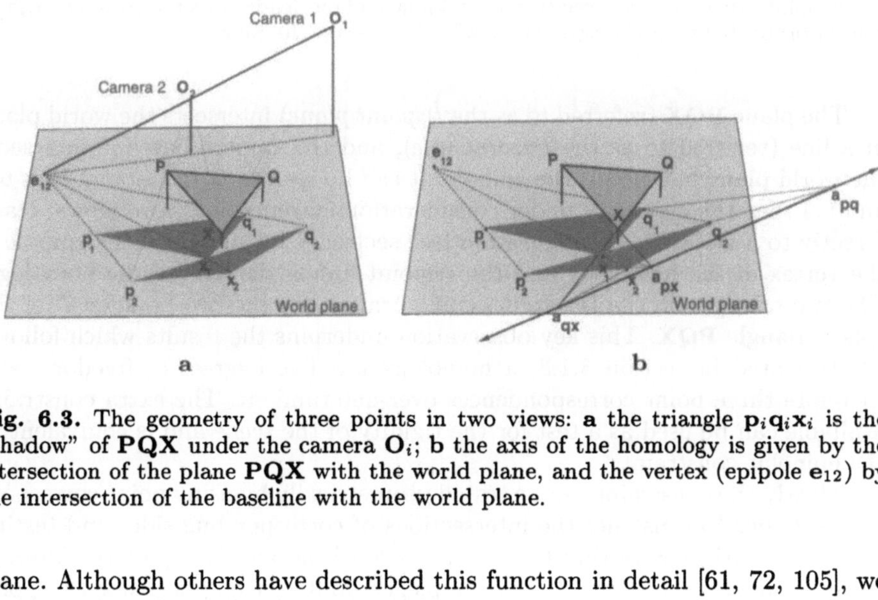

Fig. 6.3. The geometry of three points in two views: **a** the triangle $\mathbf{p}_i\mathbf{q}_i\mathbf{x}_i$ is the "shadow" of **PQX** under the camera $\mathbf{O}_i$; **b** the axis of the homology is given by the intersection of the plane **PQX** with the world plane, and the vertex (epipole $\mathbf{e}_{12}$) by the intersection of the baseline with the world plane.

plane. Although others have described this function in detail [61, 72, 105], we re-derive the relationship in section 6.3.

6.2 Geometry and duality

6.2.1 Geometry of two views

Let us consider the case of imaging seven points, four of which are coplanar from two distinct viewpoints. Each of the non-coplanar three points **P**, **Q** and **X** not on the plane gives rise to a parallax vector, which is depicted on the world plane in Fig. 6.3.

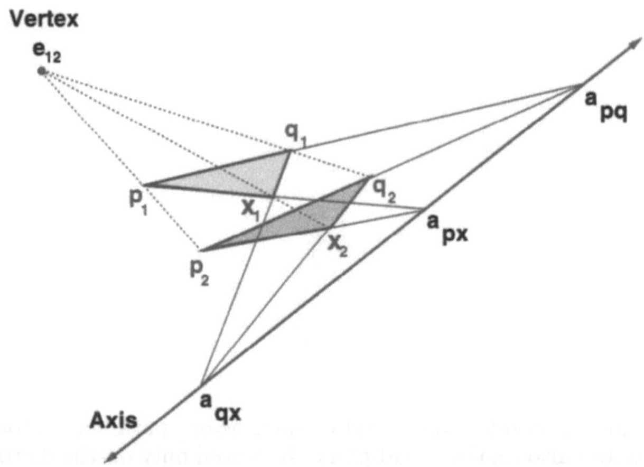

Fig. 6.4. Desargues configuration in the two view - three point (off the plane) case: three points in two views relative to a known plane leads directly to a Desargues configuration between the two "shadow" triangles on the plane.

The plane **PQX** (referred to as the *tripoint plane*) intersects the world plane in a line (referred to as the *tripoint line*), and the camera baseline intersects the world plane in a point, the epipole. It can be seen by inspection of Figs 6.3 and 6.4 that the geometry under consideration (seven points, two views) leads directly to a Desargues configuration (see section 3.1.3) in which the epipole is the vertex of the homology and the tripoint line is the axis of the homology. The two triangles in the Desargues configuration are the two "shadows" of the space triangle **PQX**. This key observation underpins the results which follow.

As stated in section 3.1.3, a homology has five degrees of freedom, and therefore three point correspondences over-determine it. The extra constraint available can be used as a test for the rigidity of the scene and is equivalent to the epipolar constraint.

Clearly the constraint can be tested geometrically by using point correspondences either to construct the intersections of corresponding sides and testing their collinearity, or testing the concurrence of the parallax vectors. Alternatively an algebraic test could, for example, compute the epipole using two point correspondences, use the epipole plus the three point correspondences to solve for a general homography, then test the homography to determine if it is a homology (a homology is represented by an invertible 3×3 matrix with two identical eigenvalues).

The geometric test has the disadvantage of requiring the construction of features which may be far removed from the measured image features themselves, while the algebraic test gives little insight into the underlying geometry.

Below we derive novel bilinear and trilinear constraints which are necessary conditions on the homology. We refer to these as *motion constraints* and they are equivalent to the epipolar constraint, but have the advantage that the computations involve only those features which can be measured directly, namely the parallax vectors.

Motion constraints.
Six points (four of which are coplanar) in two views uniquely define the epipolar geometry. Therefore the two images of one more point off the plane are constrained to lie on each other's epipolar lines.

Here we give the necessary conditions for the homology (which are therefore necessary for scene rigidity in two views) in the form of an identity involving only areas of triangles defined by three vertices, computable from the parallax vectors. Two such conditions and their symmetric forms can be determined. The first is derived from the collinearity of the epipole $\mathbf{e}_{12}$ and the corresponding points, and is bilinear in the areas. The second is derived from the collinearity of the points $\mathbf{a}_{qx}$, $\mathbf{a}_{px}$ and $\mathbf{a}_{pq}$ (Fig. 6.3**b**) and leads to a constraint which is trilinear in the areas.

Details about the derivations can be found in Appendix E. The results are summarized in table 6.2 which shows formulae for both the distinguished *plane form* (6.1–6.5) and the *image form* (6.6–6.10). In the plane form areas of triangles are computed on the distinguished plane and therefore knowledge of each world-to-image homography is required; while in the image form areas of triangles are computed directly in the first image (but either image can be used) and only the knowledge of the inter-image homography is necessary.

Violation of any of (6.1–6.10) is a clear indication that there has been non-rigid motion between the views. However if any (or all) of the points **P**, **Q** and **X** moves in its own epipolar plane, then all those equations are still satisfied and non-rigidity is not detected.

Example. An example of motion detection is shown in Fig. 6.5. Here we demonstrate two-view independent motion detection using the alternative forms of the epipolar constraint described. In this example the homography between the views induced by the world plane (the wall) has been computed using four image correspondences on each view. Two points not on the wall have then been selected and used as reference.

The bilinear motion constraint (6.6) is applied to each pair of corresponding points selected in Figs 6.5**c** and **d**. The independent motion of the person has been detected correctly (black marks). The epipolar lines have been shown for clarity in Fig. 6.5 but no explicit computation of the epipolar geometry (of the fundamental matrix) is necessary in our formulation of the motion constraint.

6.2.2 Geometry of three views

We now consider the geometry of six points, four of which are coplanar, in three views. This is the situation addressed by Irani and Anandan in [61]. The geometry is shown in Fig. 6.6.

We begin by demonstrating the duality of this case to the two view case described in section 6.2.1, and obtain a *structural constraint* directly from the measured image features. Then the trifocal tensor is derived for the three-view,

Fig. 6.5. Consistency with the epipolar geometry. **a** and **b** Left and right view of an indoor scene in which the person has moved from one sofa to another. Four pairs of corresponding points on the wall (white circles) have been used to compute the inter-image homography. Two pairs of corresponding points not on the wall plane (white boxes) have been used to compute the parallax geometry (points $\mathbf{p}$, $\mathbf{p}'$, $\mathbf{q}$ and $\mathbf{q}'$ in text). **c** and **d** Some points have been selected on the left image (white crosses, point $\mathbf{x}$ in text) and the corresponding epipolar lines shown on the right image. The corresponding points ($\mathbf{x}'$ in text) have been selected on the right image. The points which conform to the motion of the camera (they satisfy (6.6)) are marked with white crosses, the ones which undergo an independent motion (they do not satisfy (6.6)) are marked with black crosses.

	Motion Constraints	
	Plane form	
B_1 :	$A_{\mathbf{p}_1\mathbf{p}_2\mathbf{x}_2} A_{\mathbf{q}_1\mathbf{q}_2\mathbf{x}_1} = A_{\mathbf{p}_1\mathbf{p}_2\mathbf{x}_1} A_{\mathbf{q}_1\mathbf{q}_2\mathbf{x}_2}$	(6.1)
B_2 :	$A_{\mathbf{x}_1\mathbf{x}_2\mathbf{p}_2} A_{\mathbf{q}_1\mathbf{q}_2\mathbf{p}_1} = A_{\mathbf{x}_1\mathbf{x}_2\mathbf{p}_1} A_{\mathbf{q}_1\mathbf{q}_2\mathbf{p}_2}$	(6.2)
B_3 :	$A_{\mathbf{p}_1\mathbf{p}_2\mathbf{q}_2} A_{\mathbf{x}_1\mathbf{x}_2\mathbf{q}_1} = A_{\mathbf{p}_1\mathbf{p}_2\mathbf{q}_1} A_{\mathbf{x}_1\mathbf{x}_2\mathbf{q}_2}$	(6.3)
T_1 :	$A_{\mathbf{p}_1\mathbf{p}_2\mathbf{x}_1} A_{\mathbf{q}_1\mathbf{q}_2\mathbf{p}_1} A_{\mathbf{x}_1\mathbf{x}_2\mathbf{q}_1} = A_{\mathbf{p}_1\mathbf{p}_2\mathbf{q}_1} A_{\mathbf{q}_1\mathbf{q}_2\mathbf{x}_1} A_{\mathbf{x}_1\mathbf{x}_2\mathbf{p}_1}$	(6.4)
T_2 :	$A_{\mathbf{p}_1\mathbf{p}_2\mathbf{x}_2} A_{\mathbf{q}_1\mathbf{q}_2\mathbf{p}_2} A_{\mathbf{x}_1\mathbf{x}_2\mathbf{q}_2} = A_{\mathbf{p}_1\mathbf{p}_2\mathbf{q}_2} A_{\mathbf{q}_1\mathbf{q}_2\mathbf{x}_2} A_{\mathbf{x}_1\mathbf{x}_2\mathbf{p}_2}$	(6.5)
	Image form	
B_1 :	$A_{\mathbf{p}\dot{\mathbf{p}}\dot{\mathbf{x}}} A_{\mathbf{q}\dot{\mathbf{q}}\mathbf{x}} = A_{\mathbf{p}\dot{\mathbf{p}}\mathbf{x}} A_{\mathbf{q}\dot{\mathbf{q}}\dot{\mathbf{x}}}$	(6.6)
B_2 :	$A_{\mathbf{x}\dot{\mathbf{x}}\dot{\mathbf{p}}} A_{\mathbf{q}\dot{\mathbf{q}}\mathbf{p}} = A_{\mathbf{x}\dot{\mathbf{x}}\mathbf{p}} A_{\mathbf{q}\dot{\mathbf{q}}\dot{\mathbf{p}}}$	(6.7)
B_3 :	$A_{\mathbf{p}\dot{\mathbf{p}}\dot{\mathbf{q}}} A_{\mathbf{x}\dot{\mathbf{x}}\mathbf{q}} = A_{\mathbf{p}\dot{\mathbf{p}}\mathbf{q}} A_{\mathbf{x}\dot{\mathbf{x}}\dot{\mathbf{q}}}$	(6.8)
T_1 :	$A_{\mathbf{p}\dot{\mathbf{p}}\mathbf{x}} A_{\mathbf{q}\dot{\mathbf{q}}\mathbf{p}} A_{\mathbf{x}\dot{\mathbf{x}}\mathbf{q}} = A_{\mathbf{p}\dot{\mathbf{p}}\mathbf{q}} A_{\mathbf{q}\dot{\mathbf{q}}\mathbf{x}} A_{\mathbf{x}\dot{\mathbf{x}}\mathbf{p}}$	(6.9)
T_2 :	$A_{\mathbf{p}\dot{\mathbf{p}}\dot{\mathbf{x}}} A_{\mathbf{q}\dot{\mathbf{q}}\dot{\mathbf{p}}} A_{\mathbf{x}\dot{\mathbf{x}}\dot{\mathbf{q}}} = A_{\mathbf{p}\dot{\mathbf{p}}\dot{\mathbf{q}}} A_{\mathbf{q}\dot{\mathbf{q}}\dot{\mathbf{x}}} A_{\mathbf{x}\dot{\mathbf{x}}\dot{\mathbf{p}}}$	(6.10)

Table 6.2. Motion constraints: two-view bilinear (B_i) and trilinear (T_i) motion constraints equivalent to the epipolar constraint, computed on the plane and on the image. Bilinear and trilinear constraints are projective invariant.

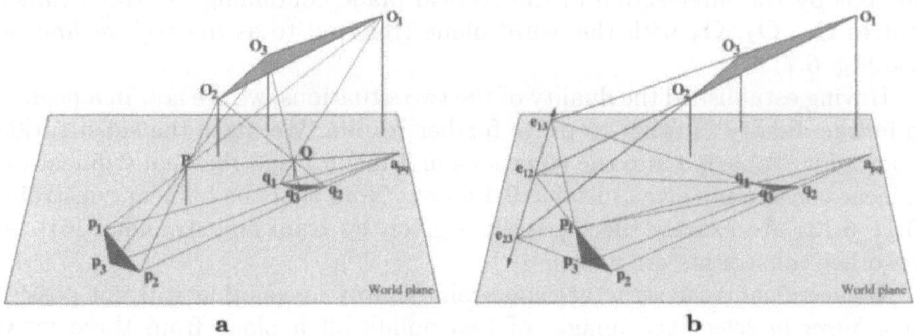

Fig. 6.6. The geometry of three views with two points off the plane. **a** The triangles $\mathbf{p}_1\mathbf{p}_2\mathbf{p}_3$ and $\mathbf{q}_1\mathbf{q}_2\mathbf{q}_3$ are "shadows" of $\mathbf{O}_1\mathbf{O}_2\mathbf{O}_3$ under the points $\mathbf{P}$ and $\mathbf{Q}$, respectively. The piercing point $\mathbf{a}_{pq}$ is the intersection of the line $< \mathbf{PQ} >$ with the world plane. **b** The three epipoles are collinear, lying on the trifocal line which is the intersection of the plane $\mathbf{O}_1\mathbf{O}_2\mathbf{O}_3$ with the world plane.

six-point (four coplanar) case. Since the trifocal tensor is over-constrained by six points, four of which are coplanar, another form of the structure constraint is also obtained.

Duality.
It is clear by inspection of Fig. 6.6 that the three view geometry is dual to that of Fig. 6.3 in which points off the plane have been exchanged for camera positions. The vertex of the homology is given by the intersection of the line $< \mathbf{PQ} >$ with the world plane (referred to as the *piercing point*, $\mathbf{a}_{pq}$), and

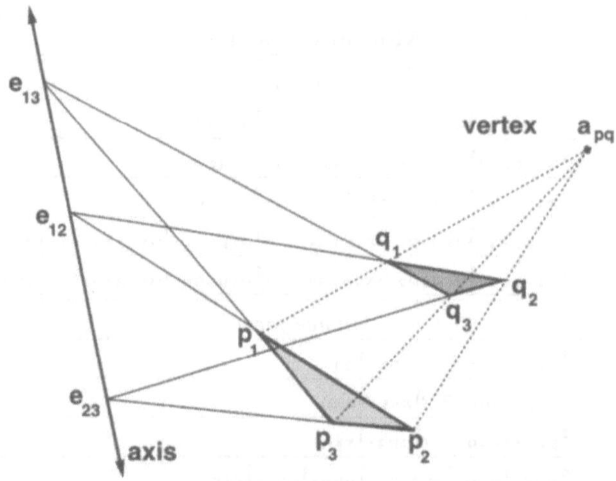

Fig. 6.7. Desargues configuration in the three-view – two-point case: the planar geometry in the three-view – two-point case is also clearly a Desargues configuration and point correspondences $\mathbf{p}_1 \leftrightarrow \mathbf{q}_1$, $\mathbf{p}_2 \leftrightarrow \mathbf{q}_2$ and $\mathbf{p}_3 \leftrightarrow \mathbf{q}_3$ are related by a homology. This situation is clearly dual to that in Fig. 6.4.

the axis by the intersection of the trifocal plane containing the three camera centres $\mathbf{O}_1$, $\mathbf{O}_2$, $\mathbf{O}_3$ with the world plane (referred to as the *trifocal line*, see also Fig. 6.7).

Having established the duality of the two situations, we are now in a position to invoke duality in order to prove further results. We make the substitutions reported in Table 6.3 into the equations in Table 6.2 and the dual trilinear and bilinear constraints given in table 6.4 follow. Note that the bilinear constraints (6.11–6.13) are exactly the constraints given by Irani and Anandan [61]. All the other constraints are new.

Geometrically the structure constraint simply means that it is not possible to *arbitrarily* select the images of two points off a plane from three views. The images of the two points are in fact constrained. For example, on the distinguished plane (see Fig. 6.6), once the projections $\mathbf{p}_1$, $\mathbf{p}_2$ and $\mathbf{p}_3$ for the point $\mathbf{P}$ and the projections $\mathbf{q}_1$ and $\mathbf{q}_2$ of the point $\mathbf{Q}$ are known, then the point $\mathbf{q}_3$ is constrained to lie on the line $< \mathbf{p}_3\mathbf{a}_{pq} >$ (*dual of the epipolar line*).

The trifocal tensor.
This section demonstrates that the trifocal tensor is uniquely determined from three views of six points, four of which are coplanar. We begin with a familiar form of the trifocal tensor (after [54]) in which we consider the image projection matrices and image point locations. We then show how the form of the tensor is simplified when we consider all geometric objects (lines, points, etc.) projected onto the distinguished plane.

Duality Substitutions						
Plane form						
2 views	$\mathbf{p}_1$	$\mathbf{p}_2$	$\mathbf{x}_1$	$\mathbf{x}_2$	$\mathbf{q}_1$	$\mathbf{q}_2$
3 views	$\mathbf{p}_1$	$\mathbf{q}_1$	$\mathbf{p}_2$	$\mathbf{q}_2$	$\mathbf{p}_3$	$\mathbf{q}_3$
Image form						
2 views	$\mathbf{p}$	$\dot{\mathbf{p}}$	$\mathbf{x}$	$\dot{\mathbf{x}}$	$\mathbf{q}$	$\dot{\mathbf{q}}$
3 views	$\mathbf{p}$	$\mathbf{q}$	$\dot{\mathbf{p}}$	$\dot{\mathbf{q}}$	$\ddot{\mathbf{p}}$	$\ddot{\mathbf{q}}$

Table 6.3. Duality substitution of symbols for both the distinguished plane form and the image form. The epipole $\mathbf{e}_{12}$ in the two-view case is dual to the piercing point $\mathbf{a}_{pq}$ in the three-view case.

Structure Constraints		
Plane form		
B_1 :	$A_{\mathbf{q}_1\mathbf{q}_2\mathbf{p}_1} A_{\mathbf{p}_2\mathbf{p}_3\mathbf{q}_3} = A_{\mathbf{p}_1\mathbf{p}_2\mathbf{q}_1} A_{\mathbf{q}_2\mathbf{q}_3\mathbf{p}_3}$	(6.11)
B_2 :	$A_{\mathbf{q}_1\mathbf{q}_2\mathbf{p}_2} A_{\mathbf{p}_1\mathbf{p}_3\mathbf{q}_3} = A_{\mathbf{p}_1\mathbf{p}_2\mathbf{q}_2} A_{\mathbf{q}_1\mathbf{q}_3\mathbf{p}_3}$	(6.12)
B_3 :	$A_{\mathbf{q}_1\mathbf{q}_3\mathbf{p}_1} A_{\mathbf{p}_2\mathbf{p}_3\mathbf{q}_2} = A_{\mathbf{p}_1\mathbf{p}_3\mathbf{q}_1} A_{\mathbf{q}_2\mathbf{q}_3\mathbf{p}_2}$	(6.13)
T_1 :	$A_{\mathbf{p}_1\mathbf{p}_2\mathbf{q}_1} A_{\mathbf{p}_1\mathbf{p}_3\mathbf{q}_3} A_{\mathbf{p}_2\mathbf{p}_3\mathbf{q}_2} = A_{\mathbf{p}_1\mathbf{p}_3\mathbf{q}_1} A_{\mathbf{p}_2\mathbf{p}_3\mathbf{q}_3} A_{\mathbf{p}_1\mathbf{p}_2\mathbf{q}_2}$	(6.14)
T_2 :	$A_{\mathbf{q}_1\mathbf{q}_2\mathbf{p}_1} A_{\mathbf{q}_1\mathbf{q}_3\mathbf{p}_3} A_{\mathbf{q}_2\mathbf{q}_3\mathbf{p}_2} = A_{\mathbf{q}_1\mathbf{q}_3\mathbf{p}_1} A_{\mathbf{q}_2\mathbf{q}_3\mathbf{p}_3} A_{\mathbf{q}_1\mathbf{q}_2\mathbf{p}_2}$	(6.15)
Image form		
B_1 :	$A_{\mathbf{q}\dot{\mathbf{q}}\mathbf{p}} A_{\dot{\mathbf{p}}\ddot{\mathbf{p}}\ddot{\mathbf{q}}} = A_{\mathbf{p}\dot{\mathbf{p}}\mathbf{q}} A_{\dot{\mathbf{q}}\ddot{\mathbf{q}}\ddot{\mathbf{p}}}$	(6.16)
B_2 :	$A_{\mathbf{q}\dot{\mathbf{q}}\dot{\mathbf{p}}} A_{\mathbf{p}\ddot{\mathbf{p}}\ddot{\mathbf{q}}} = A_{\mathbf{p}\dot{\mathbf{p}}\dot{\mathbf{q}}} A_{\mathbf{q}\ddot{\mathbf{q}}\ddot{\mathbf{p}}}$	(6.17)
B_3 :	$A_{\mathbf{q}\ddot{\mathbf{q}}\mathbf{p}} A_{\dot{\mathbf{p}}\ddot{\mathbf{p}}\dot{\mathbf{q}}} = A_{\mathbf{p}\ddot{\mathbf{p}}\mathbf{q}} A_{\dot{\mathbf{q}}\ddot{\mathbf{q}}\dot{\mathbf{p}}}$	(6.18)
T_1 :	$A_{\mathbf{p}\dot{\mathbf{p}}\mathbf{q}} A_{\mathbf{p}\ddot{\mathbf{p}}\ddot{\mathbf{q}}} A_{\dot{\mathbf{p}}\ddot{\mathbf{p}}\dot{\mathbf{q}}} = A_{\mathbf{p}\ddot{\mathbf{p}}\mathbf{q}} A_{\dot{\mathbf{p}}\ddot{\mathbf{p}}\ddot{\mathbf{q}}} A_{\mathbf{p}\dot{\mathbf{p}}\dot{\mathbf{q}}}$	(6.19)
T_2 :	$A_{\mathbf{q}\dot{\mathbf{q}}\mathbf{p}} A_{\mathbf{q}\ddot{\mathbf{q}}\ddot{\mathbf{p}}} A_{\dot{\mathbf{q}}\ddot{\mathbf{q}}\dot{\mathbf{p}}} = A_{\mathbf{q}\ddot{\mathbf{q}}\mathbf{p}} A_{\dot{\mathbf{q}}\ddot{\mathbf{q}}\ddot{\mathbf{p}}} A_{\mathbf{q}\dot{\mathbf{q}}\dot{\mathbf{p}}}$	(6.20)

Table 6.4. Structure constraints: three-view bilinear (B_i) and trilinear (T_i) structure constraints, computed on the plane and on the image. Bilinear and trilinear constraints are projective invariant.

Image form.
Suppose the homographies induced by the plane of the points are $\mathtt{H}_{12}$ and $\mathtt{H}_{13}$, such that $\mathbf{x}' = \mathtt{H}_{12}\mathbf{x}$ and $\mathbf{x}'' = \mathtt{H}_{13}\mathbf{x}$ for images of points on the plane. These homographies are computed from the images of the four coplanar points.

The images of the first camera centre in the second and third images, denoted $\mathbf{e}'$ and $\mathbf{e}''$, respectively, are the epipoles. They are determined by intersecting parallax vectors, as described in section 6.2.1, so that $\mathtt{F}_{12} = [\mathbf{e}']_\times \mathtt{H}_{12}$, and $\mathtt{F}_{13} = [\mathbf{e}'']_\times \mathtt{H}_{13}$. It is straightforward to show that the three camera projection matrices can be chosen as:

$$\mathtt{P} = [\mathtt{I} \mid \mathbf{0}], \quad \mathtt{P}' = [\mathtt{H}_{12} \mid \mathbf{e}'], \quad \mathtt{P}'' = [\mathtt{H}_{13} \mid \lambda\mathbf{e}''] \tag{6.21}$$

up to a homography of 3-space, where λ is an unknown scalar. This unknown scalar is determined by line transfer.

In fact, the line through the (non-coplanar) points $\mathbf{P}$, $\mathbf{Q}$ is imaged as $\mathbf{l} = \mathbf{p}\times\mathbf{q}$, $\mathbf{l}' = \mathbf{p}'\times\mathbf{q}'$, $\mathbf{l}'' = \mathbf{p}''\times\mathbf{q}''$ in the first, second and third views, respectively. It is then straightforward to show that lines transfer as follows (alternatively point transfer may be considered):

$$\mathbf{l} = \lambda(\mathbf{e}''\cdot\mathbf{l}'')\mathtt{H}_{12}{}^{\top}\mathbf{l}' - (\mathbf{e}'\cdot\mathbf{l}')\mathtt{H}_{13}{}^{\top}\mathbf{l}'' \tag{6.22}$$

The scalar λ is the only unknown in this equation. It is determined by taking the vector product with $\mathbf{l}$

$$\lambda(\mathbf{e}''\cdot\mathbf{l}'')\mathbf{l}\times(\mathtt{H}_{12}{}^{\top}\mathbf{l}') = (\mathbf{e}'\cdot\mathbf{l}')\mathbf{l}\times(\mathtt{H}_{13}{}^{\top}\mathbf{l}'') \tag{6.23}$$

This provides two equations in the one unknown λ and so we can solve uniquely for the trifocal tensor and obtain one further constraint, namely the rigidity condition that the imaged intersection of the line through $\mathbf{P},\mathbf{Q}$ is the same when computed from views one and two ($\mathbf{l}\times(\mathtt{H}_{12}{}^{\top}\mathbf{l}')$) as from views one and three ($\mathbf{l}\times(\mathtt{H}_{13}{}^{\top}\mathbf{l}'')$). This is yet another form of the constraints (6.11–6.20). The scale factor λ is obtained by normalizing both sides of (6.23):

$$\lambda = \frac{||(\mathbf{e}'\cdot\mathbf{l}')\mathbf{l}\times(\mathtt{H}_{13}{}^{\top}\mathbf{l}'')||}{||(\mathbf{e}''\cdot\mathbf{l}'')\mathbf{l}\times(\mathtt{H}_{12}{}^{\top}\mathbf{l}')||} \tag{6.24}$$

Distinguished plane form.
On the distinguished plane $\mathtt{H}_{12} = \mathtt{H}_{13} = \mathtt{I}$, so the equivalent of (6.22) for point transfer is:

$$\mathbf{x}_3 = \lambda\mathbf{e}_{13}(\mathbf{l}_2\cdot\mathbf{x}_1) - (\mathbf{e}_{12}\cdot\mathbf{l}_2)\mathbf{x}_1 \tag{6.25}$$

where $\mathbf{x}_1$, $\mathbf{x}_2$, $\mathbf{x}_3$ are the distinguished plane images of a general three-dimensional point $\mathbf{X}$, and $\mathbf{l}_2$ is any line through $\mathbf{x}_2$. This equation depends only on the positions of the epipoles on the distinguished plane, with all dependence on camera internals and relative rotations having been factored out into the image-to-plane homographies.

Additionally the projection matrices have the very simple form:

$$\mathtt{P}_1 = [\mathtt{I} \mid \mathbf{0}], \quad \mathtt{P}_2 = [\mathtt{I} \mid \mathbf{e}_{12}], \quad \mathtt{P}_3 = [\mathtt{I} \mid \lambda\mathbf{e}_{13}] \tag{6.26}$$

Hence, representing a general three-dimensional point as $\mathbf{X} = \begin{pmatrix} x_1 \\ \rho \end{pmatrix}$, we determine the distinguished plane images to be:

$$\mathbf{x}_1 = \mathtt{P}_1\mathbf{X}, \quad \mathbf{x}_2 = \mathtt{P}_2\mathbf{X} = \mathbf{x}_1 + \rho\mathbf{e}_{12}, \quad \mathbf{x}_3 = \mathtt{P}_3\mathbf{X} = \mathbf{x}_1 + \rho\lambda\mathbf{e}_{13} \tag{6.27}$$

We now give an interpretation of ρ and λ on the distinguished plane (see Fig. 6.8).

The ratio λ depends only on the camera centres, not on the points, and can be determined as $\lambda = d(\mathbf{e}_{12},\mathbf{e}_{23})/d(\mathbf{e}_{13},\mathbf{e}_{23})$, where $d()$ is the distance between the points on the distinguished plane. The parameter ρ is the relative affine invariant of Shashua [110], and is related to the point depth. On the distinguished plane it is obtained as $\rho = d(\mathbf{x}_2,\mathbf{x}_1)/d(\mathbf{x}_2,\mathbf{e}_{12})$.

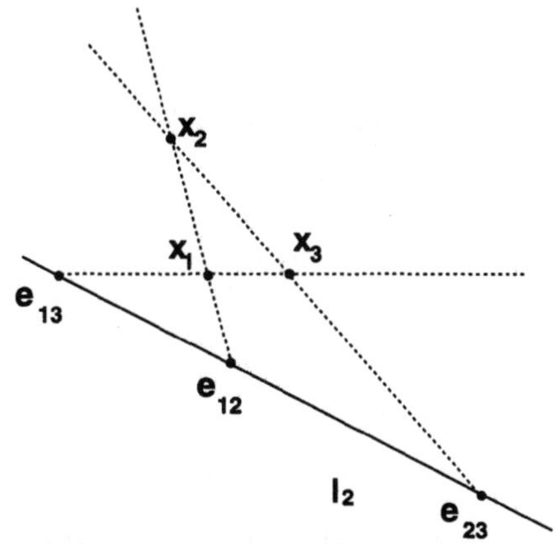

Fig. 6.8. Point transfer in three views: the ratios of the distances between $\mathbf{x}_1$, $\mathbf{x}_2$ and $\mathbf{e}_{12}$ and the three epipoles define the transfer of the point $\mathbf{x}_1$ to $\mathbf{x}_3$. The scalars λ and ρ (see text) can be measured as ratios of point distances as: $\lambda = d(\mathbf{e}_{12}, \mathbf{e}_{23})/d(\mathbf{e}_{13}, \mathbf{e}_{23})$ and $\rho = d(\mathbf{x}_2, \mathbf{x}_1)/d(\mathbf{x}_2, \mathbf{e}_{12})$.

So point transfer using the trifocal tensor simply involves computing the ratio ρ from $\mathbf{x}_1$, $\mathbf{x}_2$ and $\mathbf{e}_{12}$ and employing λ to define the transferred point $\mathbf{x}_3$ on the line between $\mathbf{e}_{13}$ and $\mathbf{x}_1$ as $\mathbf{x}_3 = \mathbf{x}_1 + \rho\lambda\mathbf{e}_{13}$ in (6.27). This is identical to the point transfer of (6.25), as can be seen by considering similar triangles in Fig. 6.8.

Trifocal transfer in the case of collinear camera centres is not degenerate. In the case where the three camera centres are collinear then the three epipoles $\mathbf{e}_{13}$, $\mathbf{e}_{23}$ and $\mathbf{e}_{12}$ coincide and the epipolar transfer would fail. The trifocal transfer, in contrast, does not present such a degeneracy. It is still well defined in the case of three concurrent (or almost concurrent) epipoles. In fact, the ratio λ is defined and can be obtained using the distinguished plane equivalent of (6.24).

6.3 Scene reconstruction

In this section we discuss a number of useful structural computations which can be achieved using ratios of areas. This requires *affine* measurements on the world-plane, which can be obtained either from four world plane points known up to an affinity (and hence the image-to-world plane homographies, this is *distinguished plane form case*), or from the inter-image homography between two views and the vanishing line of the world plane in either image (this is the *image form case*). In either case we obtain results for the scene structure without resorting first to compute the epipolar geometry.

A significant novel aspect of the formulae given in sections 6.3.1 and 6.3.2 is that measurements can be computed without knowing the vanishing point

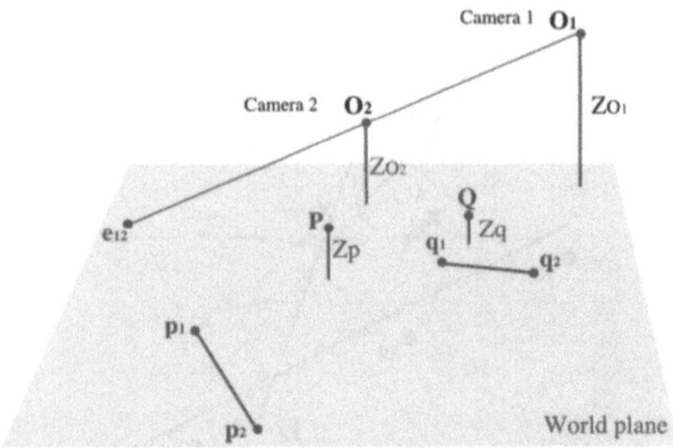

Fig. 6.9. Parallax geometry of two points **P** and **Q** off the plane.

for the reference direction and without seeing the base point in the image, in contrast to the single-view metrology approach described in Chapter 5.

We begin by re-deriving the basic parallax relationship for the cases where the parallax is measured on the distinguished world plane (6.30) and on the image (6.31).

Distinguished plane form.
Considering Fig. 6.9 and writing $\mathbf{p}_1 = \mathbf{O}_1 + \frac{Z_{O_1}}{Z_{O_1}-Z_p}(\mathbf{P} - \mathbf{O}_1)$, $\mathbf{p}_2 = \mathbf{O}_2 + \frac{Z_{O_2}}{Z_{O_2}-Z_p}(\mathbf{P} - \mathbf{O}_2)$ and $\mathbf{e}_{12} = \mathbf{O}_2 + \frac{Z_{O_2}}{Z_{O_2}-Z_p}(\mathbf{O}_1 - \mathbf{O}_2)$. Then eliminating the camera centres $\mathbf{O}_1$ and $\mathbf{O}_2$ yields:

$$\boldsymbol{\mu}_p = \frac{Z_p}{Z_{O_2} - Z_p} \frac{\Delta_{Z_O}}{Z_{O_1}} (\mathbf{p}_1 - \mathbf{e}_{12}) \tag{6.28}$$

where $\boldsymbol{\mu}_p = \mathbf{p}_2 - \mathbf{p}_1$ is the planar parallax vector and $\Delta_{Z_O} = Z_{O_1} - Z_{O_2}$ is the component of camera translation towards the plane.

Let γ be the ratio of the distance of a point to the plane and the point to the first camera (measured in the same direction), i.e. $\gamma_p = \frac{Z_p}{Z_{O_1}-Z_p}$ and $\gamma_q = \frac{Z_q}{Z_{O_1}-Z_q}$, then combining the basic parallax equation (6.28) for two points **P** and **Q** gives:

$$\gamma_q \boldsymbol{\mu}_p - \gamma_p \boldsymbol{\mu}_q = \gamma_p \gamma_q \frac{\Delta_{Z_O}}{Z_{O_2}} (\mathbf{p}_2 - \mathbf{q}_2) \tag{6.29}$$

Finally, taking the cross-product of both sides of the equation with $\mathbf{p}_2 - \mathbf{q}_2$ and taking magnitudes yields an expression for $\frac{\gamma_q}{\gamma_p}$ as a ratio of areas (see Fig. 6.10) of the form

$$\frac{\gamma_q}{\gamma_p} = \frac{|\boldsymbol{\mu}_q \times (\mathbf{p}_2 - \mathbf{q}_2)|}{|\boldsymbol{\mu}_p \times (\mathbf{p}_2 - \mathbf{q}_2)|} = \frac{A_{\mathbf{q}_1\mathbf{q}_2\mathbf{p}_2}}{A_{\mathbf{p}_1\mathbf{p}_2\mathbf{q}_2}} \tag{6.30}$$

This ratio is computable solely from the parallax measurements, and, being a ratio of areas, is clearly affine invariant.

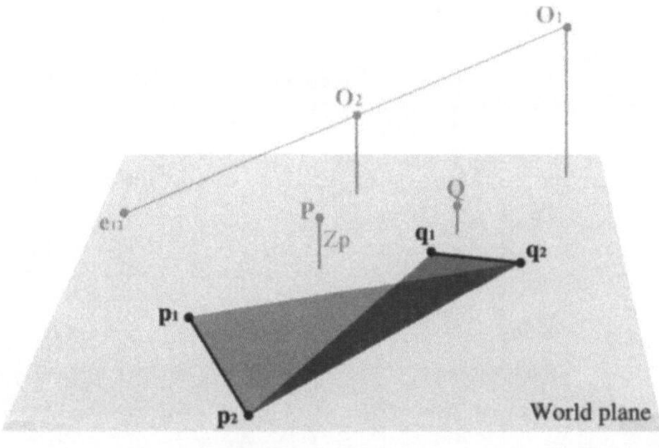

Fig. 6.10. The relative structure $\frac{\gamma_q}{\gamma_p}$ can be expressed as a ratio of areas on the world plane as in (6.30).

Image form.
In the case where the image-to-world homographies are unknown, since $\dot{\mathbf{p}} = \mathtt{H}_{21}\mathbf{p}'$ and $\dot{\mathbf{q}} = \mathtt{H}_{21}\mathbf{q}'$, equation (6.30) becomes (see section E.2 in Appendix E, for details):

$$\frac{\gamma_q}{\gamma_p} = \frac{A_{\mathbf{q}\dot{\mathbf{q}}\dot{\mathbf{p}}}(\mathbf{l}\cdot\mathbf{p})}{A_{\mathbf{p}\dot{\mathbf{p}}\dot{\mathbf{q}}}(\mathbf{l}\cdot\mathbf{q})} \tag{6.31}$$

with $\mathbf{l}$ the vanishing line of the reference plane in the first image. A similar formula can be obtained with reference to the second image replacing the first one.

This derivation is equivalent to Irani and Anandan's construction [61], but note that in our formulation only affine constructs have been used (no orthogonality has been assumed and the formulae are homogeneous). Furthermore, thanks to the affine invariance of ratios of areas if the vanishing line $\mathbf{l}$ of the reference plane is known in either image, then the ratio (6.30) can be computed from image measurements only, thus obtaining (6.31).

These results are derived for the two-view, seven-point case. However because of the fundamental duality proved in section 6.2.2, they are equally valid (with appropriate symbol substitutions, see Table 6.3) in the three-view, six-point case. For example (6.38) can be used to compute the height of a third point given two other known heights in the two view, seven point case; dually, in the three view, six point case, it can be used to obtain the height of a third camera given the other two camera heights.

6.3.1 Determining the position of the cameras

This section presents formulae to compute the position of the cameras from a pair of images, with respect to the reference plane.

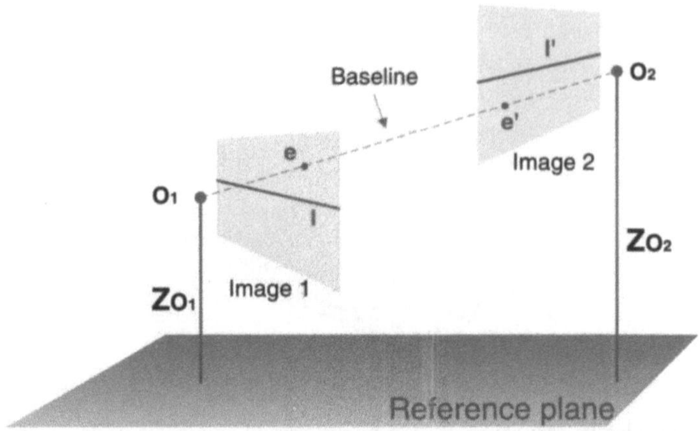

Fig. 6.11. The order in the distances of the two cameras from the common reference plane can be estimated from a quick analysis of the relative location of epipoles and vanishing lines.

Ordinal distances.

To begin, we show that the position of the epipoles $\mathbf{e}$ and $\mathbf{e}'$ and the plane vanishing lines $\mathbf{l}$ and $\mathbf{l}'$ in the two images provide useful information about the relative location of the two cameras. Given the vanishing line $\mathbf{l}$ of the reference plane on the first image, its corresponding vanishing line $\mathbf{l}'$ in the second image is computed as: $\mathbf{l}' = \mathtt{H}_{12}^{-\top}\mathbf{l}$ (where $\mathbf{x}' = \mathtt{H}_{12}\mathbf{x}$ for images of points on the reference plane).

As noted in section 5.1 the vanishing line of a plane partitions the scene points into the ones which are closer to the plane than the camera centre and the ones which are farther (see Fig. 5.2b). This simple consideration, generalized in the two view case, leads to the following result.

If we denote the epipoles in the two images with $\mathbf{e}$ and $\mathbf{e}'$, respectively, and the two vanishing lines of the reference plane with $\mathbf{l}$ and $\mathbf{l}'$ respectively, then:

- $\mathbf{e} \cdot \mathbf{l} = 0 \iff \mathbf{e}' \cdot \mathbf{l}' = 0$ and in this case the two cameras have the same distance from the distinguished plane $\boldsymbol{\pi}$, $Z_{O_2} = Z_{O_1}$;
- if $\mathbf{e} \cdot \mathbf{l} \neq 0$, one camera is higher than the other; for instance, if $\mathbf{e}$ lies above the vanishing line then $\mathbf{e}'$ lies below and $Z_{O_2} > Z_{O_1}$ (see Fig. 6.11). The reverse is also true.

Therefore we obtain ordering for the distances of the cameras from the reference plane $\boldsymbol{\pi}$.

Affine distances.

If two reference points are seen in both images then affine distances of the cameras can be computed (cf. Fig. 6.9. The vanishing point of the measuring direction is not required. The following theorem holds:

Theorem 6.3.1. *Given the planar parallax vectors of two reference world points* $\mathbf{P}$ *and* $\mathbf{Q}$*, the ratio of distances of these points from the world plane*

$\frac{Z_p}{Z_q}$ *(measured in the same but arbitrary direction, with $Z_p \neq Z_q$), and affine measurements on the world plane; then the affine distance of either camera to the world plane (measured in the same direction as Z_p and Z_q) can be computed.*

Distinguished plane form. Rearranging (6.30) provides the following expressions for the distance of the cameras from the plane:

$$\frac{Z_{O_1}}{Z_p} = \frac{A_{\mathbf{p_1p_2q_2}} - A_{\mathbf{q_1q_2p_2}}}{A_{\mathbf{p_1p_2q_2}} - \frac{Z_p}{Z_q}A_{\mathbf{q_1q_2p_2}}} \qquad (6.32)$$

$$\frac{Z_{O_2}}{Z_p} = \frac{A_{\mathbf{p_1p_2q_1}} - A_{\mathbf{q_1q_2p_1}}}{A_{\mathbf{p_1p_2q_1}} - \frac{Z_p}{Z_q}A_{\mathbf{q_1q_2p_1}}}$$

Image form. Taking into account (6.31), the previous equations can be computed directly in the image as:

$$\frac{Z_{O_1}}{Z_p} = \frac{(\mathbf{l}\cdot\mathbf{q})A_{\mathbf{p\dot{p}\dot{q}}} - (\mathbf{l}\cdot\mathbf{p})A_{\mathbf{q\dot{q}\dot{p}}}}{(\mathbf{l}\cdot\mathbf{q})A_{\mathbf{p\dot{p}\dot{q}}} - \frac{Z_p}{Z_q}(\mathbf{l}\cdot\mathbf{p})A_{\mathbf{q\dot{q}\dot{p}}}} \qquad (6.33)$$

$$\frac{Z_{O_2}}{Z_p} = \frac{(\mathbf{l}'\cdot\mathbf{q}')A_{\mathbf{p\dot{p}q}} - (\mathbf{l}'\cdot\mathbf{p}')A_{\mathbf{q\dot{q}p}}}{(\mathbf{l}'\cdot\mathbf{q}')A_{\mathbf{p\dot{p}q}} - \frac{Z_p}{Z_q}(\mathbf{l}'\cdot\mathbf{p}')A_{\mathbf{q\dot{q}p}}}$$

Since $\frac{Z_{O_1}}{Z_p}$ and $\frac{Z_{O_2}}{Z_p}$ scale linearly with the unknown Z_p, we have obtained affine structure.

□

Euclidean distances.
An immediate corollary follows from the previous theorem. In fact, if the values of the reference distances Z_p and Z_q are known, then the above measurements can be upgraded to Euclidean.

Distinguished plane form. From (6.32) we obtain:

$$Z_{O_1} = \frac{Z_pZ_q(A_{\mathbf{p_1p_2q_2}} - A_{\mathbf{q_1q_2p_2}})}{Z_qA_{\mathbf{p_1p_2q_2}} - Z_pA_{\mathbf{q_1q_2p_2}}} \qquad (6.34)$$

$$Z_{O_2} = \frac{Z_pZ_q(A_{\mathbf{p_1p_2q_1}} - A_{\mathbf{q_1q_2p_1}})}{Z_qA_{\mathbf{p_1p_2q_1}} - Z_pA_{\mathbf{q_1q_2p_1}}}$$

Image form. And from (6.33):

$$Z_{O_1} = \frac{Z_pZ_q\left[(\mathbf{l}\cdot\mathbf{q})A_{\mathbf{p\dot{p}\dot{q}}} - (\mathbf{l}\cdot\mathbf{p})A_{\mathbf{q\dot{q}\dot{p}}}\right]}{Z_q(\mathbf{l}\cdot\mathbf{q})A_{\mathbf{p\dot{p}\dot{q}}} - Z_p(\mathbf{l}\cdot\mathbf{p})A_{\mathbf{q\dot{q}\dot{p}}}} \qquad (6.35)$$

$$Z_{O_2} = \frac{Z_pZ_q\left[(\mathbf{l}'\cdot\mathbf{q}')A_{\mathbf{p\dot{p}q}} - (\mathbf{l}'\cdot\mathbf{p}')A_{\mathbf{q\dot{q}p}}\right]}{Z_q(\mathbf{l}'\cdot\mathbf{q}')A_{\mathbf{p\dot{p}q}} - Z_p(\mathbf{l}'\cdot\mathbf{p}')A_{\mathbf{q\dot{q}p}}}$$

Example. In the example in Fig. 6.16 we have used (6.34) to compute the heights of the cameras above the floor as 566 cm and 586 cm (left and right views respectively).

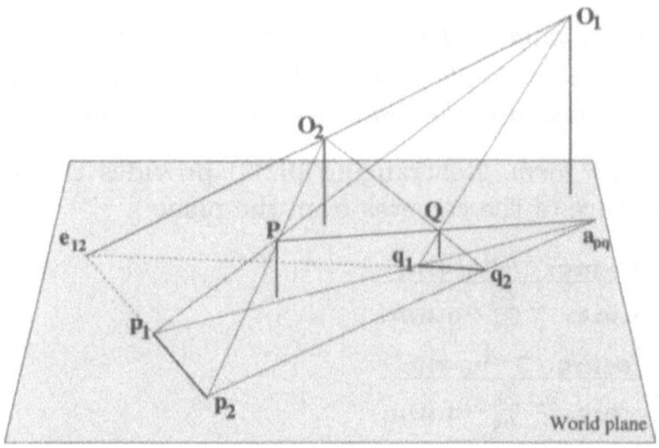

Fig. 6.12. Computing ordinal heights of points from parallax vectors and piercing point: if $d(\mathbf{p}_1\mathbf{a}_{pq}) > d(\mathbf{q}_1\mathbf{a}_{pq})$ then $Z_p > Z_q$ and viceversa (cf. Fig. 6.9).

6.3.2 Distances of points from planes

Ordinal distances.
If two points off the distinguished plane **P** and **Q** are viewed in two images, then it is possible to obtain ordering for their distances from the plane. This is obtained by comparing the locations of the projections of the points onto the plane and the piercing point. By inspection of Fig. 6.12 if $d(\mathbf{p}_1, \mathbf{a}_{pq}) > d(\mathbf{q}_1, \mathbf{a}_{pq})$ then $Z_p > Z_q$ and viceversa. Similar computation can be performed directly on the images.

Affine distances.
This section demonstrates that affine height of points can be computed from two references. As before, the vanishing point for the reference direction is not required. The following theorem holds (cf. Fig. 6.3**a**):

Theorem 6.3.2. *Given the parallax vectors of three world points* **P**, **Q** *and* **X**, *the ratio of the distances of two of these points* $\frac{Z_p}{Z_q}$ *from the world plane (measured in the same but arbitrary direction, with* $Z_p \neq Z_q$*), and affine measurements on the world plane; then the affine distance of the third point* **X** *from the plane (measured in the same direction as* Z_p *and* Z_q*) can be computed.*

Distinguished plane form. From (6.30) we have that $\frac{\gamma_x}{\gamma_p} = \frac{A_{\mathtt{x_1x_2p_1}}}{A_{\mathtt{p_1p_2x_1}}}$ and $\frac{\gamma_x}{\gamma_q} = \frac{A_{\mathtt{x_1x_2q_1}}}{A_{\mathtt{q_1q_2x_1}}}$. After eliminating the camera distance between these equations we obtain an expression for the ratio Z_x/Z_p where Z_x is the distance of the point **X** from the plane (see Fig. 6.3):

$$\frac{Z_x}{Z_p} = \frac{|\mathtt{M}_1|}{|\mathtt{M}_2|} \tag{6.36}$$

where:

$$\mathtt{M}_1 = \begin{bmatrix} A_{\mathtt{x_1x_2q_1}} & A_{\mathtt{q_1q_2x_1}} \\ A_{\mathtt{x_1x_2p_1}} & A_{\mathtt{p_1p_2x_1}} \end{bmatrix}, \quad \mathtt{M}_2 = \begin{bmatrix} A_{\mathtt{x_1x_2q_1}} & A_{\mathtt{q_1q_2x_1}} & A_{\mathtt{q_1q_2x_1}} \\ Z_p/Z_q & 0 & -1 \\ A_{\mathtt{p_1p_2x_1}} & A_{\mathtt{p_1p_2x_1}} & A_{\mathtt{x_1x_2p_1}} \end{bmatrix}$$

Image form. The ratio $\frac{Z_x}{Z_p}$ in (6.36) can be computed directly in the image as:

$$\frac{Z_x}{Z_p} = \frac{|\tilde{\mathtt{M}}_1|}{|\tilde{\mathtt{M}}_2|} \tag{6.37}$$

where:

$$\tilde{\mathtt{M}}_1 = \begin{bmatrix} (\mathbf{l}\cdot\mathbf{q})A_{\mathbf{x}\dot{\mathbf{x}}\dot{\mathbf{q}}} & A_{\mathbf{q}\dot{\mathbf{q}}\dot{\mathbf{x}}} \\ (\mathbf{l}\cdot\mathbf{p})A_{\mathbf{x}\dot{\mathbf{x}}\dot{\mathbf{p}}} & A_{\mathbf{p}\dot{\mathbf{p}}\dot{\mathbf{x}}} \end{bmatrix}, \quad \tilde{\mathtt{M}}_2 = \begin{bmatrix} (\mathbf{l}\cdot\mathbf{q})A_{\mathbf{x}\dot{\mathbf{x}}\dot{\mathbf{q}}} & A_{\mathbf{q}\dot{\mathbf{q}}\dot{\mathbf{x}}} & (\mathbf{l}\cdot\mathbf{x})A_{\mathbf{q}\dot{\mathbf{q}}\dot{\mathbf{x}}} \\ Z_p/Z_q & 0 & -1 \\ (\mathbf{l}\cdot\mathbf{x})A_{\mathbf{p}\dot{\mathbf{p}}\dot{\mathbf{x}}} & A_{\mathbf{p}\dot{\mathbf{p}}\dot{\mathbf{x}}} & (\mathbf{l}\cdot\mathbf{p})A_{\mathbf{x}\dot{\mathbf{x}}\dot{\mathbf{p}}} \end{bmatrix}$$

thus affine structure is recovered.

□

The fact that the above measurements are expressed in terms of ratios of determinants is a clear indication that they can be obtained by employing Grassman-Cayley algebra [42, 124].

Euclidean distances.

As for the camera, also in this case can the formulae above be upgraded to Euclidean if the absolute distances of the two reference points Z_p and Z_q are known.

Distinguished plane form. From (6.36) we obtain:

$$Z_x = \frac{|\mathtt{M}_1|}{|\mathtt{M}_2|} \tag{6.38}$$

where:

$$\mathtt{M}_1 = Z_p Z_q \begin{bmatrix} A_{\mathbf{x}_1\mathbf{x}_2\mathbf{q}_1} & A_{\mathbf{q}_1\mathbf{q}_2\mathbf{x}_1} \\ A_{\mathbf{x}_1\mathbf{x}_2\mathbf{p}_1} & A_{\mathbf{p}_1\mathbf{p}_2\mathbf{x}_1} \end{bmatrix}, \quad \mathtt{M}_2 = \begin{bmatrix} A_{\mathbf{x}_1\mathbf{x}_2\mathbf{q}_1} & A_{\mathbf{q}_1\mathbf{q}_2\mathbf{x}_1} & A_{\mathbf{q}_1\mathbf{q}_2\mathbf{x}_1} \\ Z_p & 0 & -Z_q \\ A_{\mathbf{p}_1\mathbf{p}_2\mathbf{x}_1} & A_{\mathbf{p}_1\mathbf{p}_2\mathbf{x}_1} & A_{\mathbf{x}_1\mathbf{x}_2\mathbf{p}_1} \end{bmatrix}$$

Image form. And from (6.37):

$$Z_x = \frac{|\tilde{\mathtt{M}}_1|}{|\tilde{\mathtt{M}}_2|} \tag{6.39}$$

where:

$$\tilde{\mathtt{M}}_1 = Z_p Z_q \begin{bmatrix} (\mathbf{l}\cdot\mathbf{q})A_{\mathbf{x}\dot{\mathbf{x}}\dot{\mathbf{q}}} & A_{\mathbf{q}\dot{\mathbf{q}}\dot{\mathbf{x}}} \\ (\mathbf{l}\cdot\mathbf{p})A_{\mathbf{x}\dot{\mathbf{x}}\dot{\mathbf{p}}} & A_{\mathbf{p}\dot{\mathbf{p}}\dot{\mathbf{x}}} \end{bmatrix}, \quad \tilde{\mathtt{M}}_2 = \begin{bmatrix} (\mathbf{l}\cdot\mathbf{q})A_{\mathbf{x}\dot{\mathbf{x}}\dot{\mathbf{q}}} & A_{\mathbf{q}\dot{\mathbf{q}}\dot{\mathbf{x}}} & (\mathbf{l}\cdot\mathbf{x})A_{\mathbf{q}\dot{\mathbf{q}}\dot{\mathbf{x}}} \\ Z_p & 0 & -Z_q \\ (\mathbf{l}\cdot\mathbf{x})A_{\mathbf{p}\dot{\mathbf{p}}\dot{\mathbf{x}}} & A_{\mathbf{p}\dot{\mathbf{p}}\dot{\mathbf{x}}} & (\mathbf{l}\cdot\mathbf{p})A_{\mathbf{x}\dot{\mathbf{x}}\dot{\mathbf{p}}} \end{bmatrix}$$

Derivations of these formulae are shown in Appendix E.

Ambiguity in structure computation.

As mentioned above, the distance Z_x is computed along the direction of the two reference distances Z_p and Z_q (parallel to each other in any fixed direction not on the reference plane, Fig. 6.13). But the reference direction is not known (its vanishing point is not needed in (6.38, 6.39)) and the position of the projection of the point **X** on the plane (its X and Y coordinates) cannot be completely determined.

It can be proven, though, that the X, Y position of the world point **X** can be retrieved up to a one-parameter family. In fact, the X and Y coordinates are related by a linear function. Furthermore, knowledge of the X or Y coordinate of at least one of the two reference points (**P** or **Q**) completely constrains the X, Y position of the point **X**. Similar analysis applies to camera computation.

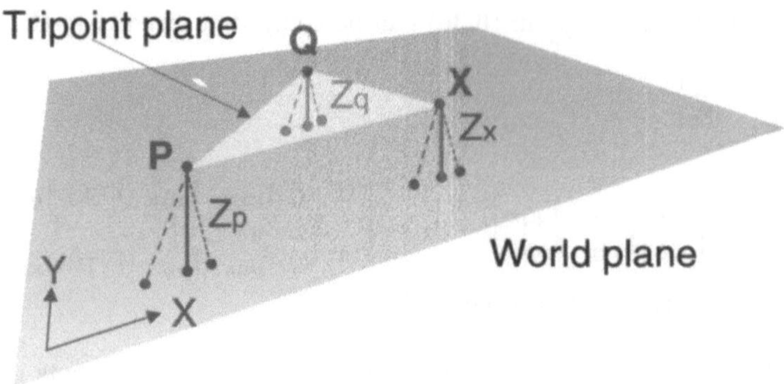

Fig. 6.13. Ambiguity in the reconstruction using a plane-plus-parallax approach: the distance Z_x of the point $\mathbf{X}$ from the world plane is computed using (6.38) or (6.39) but its X, Y coordinates are not defined since the reference direction (off the plane) is not defined. The only constraint is that Z_x, Z_p and Z_q are all measured along the same direction.

Degeneracy.
Note that in the case where the reference heights coincide ($Z_p = Z_q$) then the matrices above are singular and (6.32–6.39) are degenerate. However this situation can be avoided in practice.

The degeneracy is understood in terms of the geometry as follows. In general we obtain a projective structure (since $\mathbf{F}$ is determined uniquely). If in addition $Z_p \neq Z_q$ (Fig. 6.14**a**), the line $< \mathbf{PQ} >$ intersects the plane at infinity π_∞ in the vanishing point $\mathbf{v}_{pq}$ which can be identified in both images since the four aligned points $\mathbf{p}$, $\mathbf{q}$, $\mathbf{a}_{pq}$ and $\mathbf{v}_{pq}$ define a cross-ratio and the ratio $\frac{Z_p}{Z_q}$ is known (Fig. 6.15). The point $\mathbf{v}_{pq}$ and the vanishing line of the world plane, $\mathbf{l}$, determine π_∞, hence we can obtain affine structure. When $Z_p = Z_q$ (Fig. 6.14**b**) then the line $< \mathbf{PQ} >$ intersects π_∞ on $\mathbf{l}$ and so no additional information about π_∞ is obtained; it is determined only up to a one-parameter family (the pencil of planes with $\mathbf{l}$ as its axis).

Examples.
Figure 6.16 shows an example in which point distances from the floor have been estimated. Two views of a dining hall are taken from above. The distance between the two cameras is about 5 m. The floor tiling and the perpendicular heights of two points in the hall were measured by hand with a tape measure and used as a reference. Correspondences (more than four corresponding points) on the patterned floor tiling have been used to compute the inter-image homography induced by the ground plane. Parallel lines identified by the floor pattern have been employed to compute the vanishing line of the floor in the left image (the horizon). The distances of several points have, then, been computed with the parallax- based method described above and shown in the figure. Note, for instance, that the step was computed to be 11.4cm high and measured by hand as 11.0 cm. Notice also how close the heights of the two chandeliers are to each other (322.1 cm and 326.0 cm). A second example is shown in Fig. 6.19.

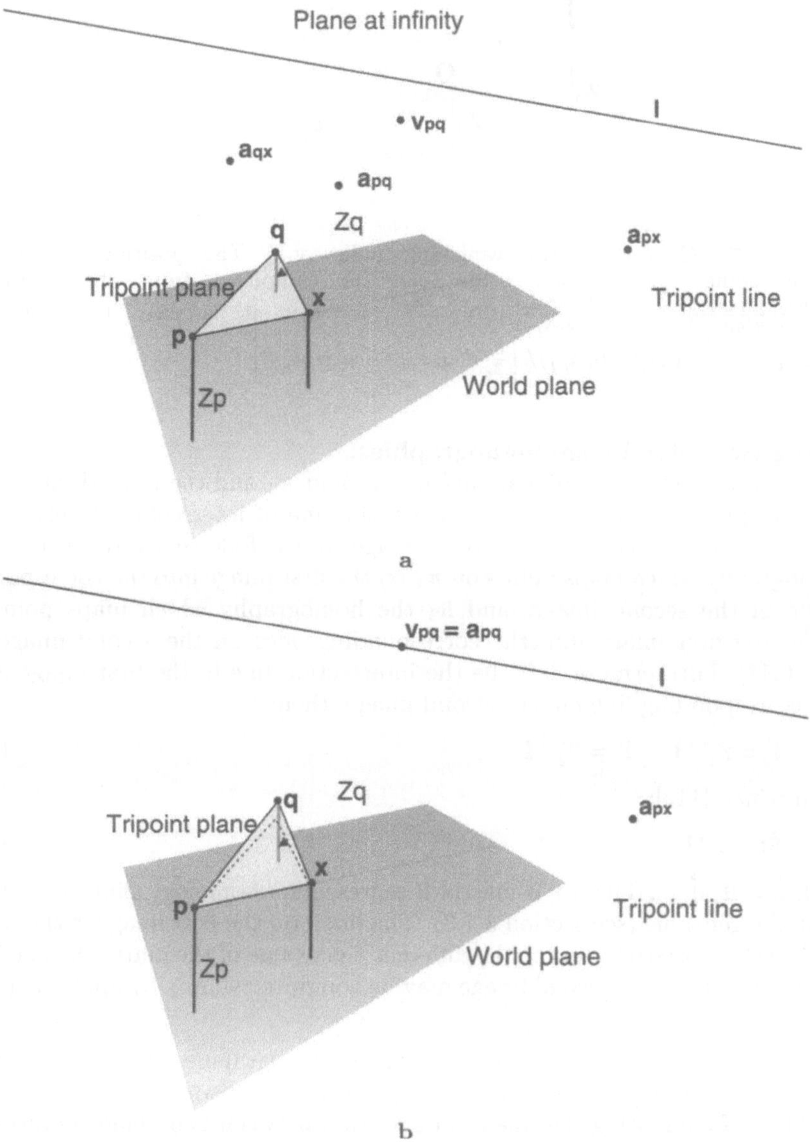

Fig. 6.14. The plane-plus-parallax approach is degenerate if the two reference distances coincide: **a** if $Z_p \neq Z_q$ the point $\mathbf{v}_{pq}$ (vanishing point for the direction defined by the line $< \mathbf{PQ} >$) does not lie on the plane vanishing line **l**, then the plane at infinity is defined. **b** if $Z_p = Z_q$ the point $\mathbf{v}_{pq}$ lies on the plane vanishing line **l** and the plane at infinity is not defined.

6.3.3 Intersection between two planar surfaces

Given two images of two planar surfaces, the line of intersection between them can be computed in different ways. In this section we show the advantages of using a plane-plus-parallax approach over other methods.

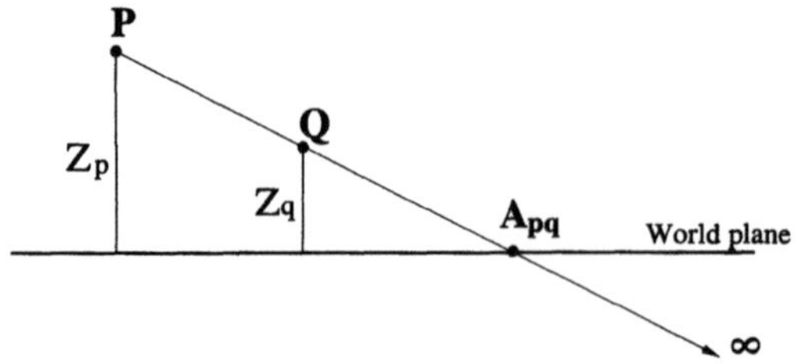

Fig. 6.15. Computing the vanishing point $\mathbf{v}_{pq}$. The position of the vanishing point $\mathbf{v}_{pq}$ in Fig. 6.14**a** can be computed from the cross-ratio: $\frac{d(\mathbf{p},\mathbf{a}_{pq})}{d(\mathbf{q},\mathbf{a}_{pq})}\frac{d(\mathbf{p},\mathbf{v}_{pq})}{d(\mathbf{q},\mathbf{v}_{pq})} = \frac{d(\mathbf{P},\mathbf{A}_{pq})}{d(\mathbf{Q},\mathbf{A}_{pq})} = \frac{Z_p}{Z_q}$. In fact it is easy to prove that $d(\mathbf{q},\mathbf{v}_{pq}) = d(\mathbf{p},\mathbf{q})d(\mathbf{p},\mathbf{a}_{pq})/\left[\frac{Z_p}{Z_q}d(\mathbf{q},\mathbf{a}_{pq}) - d(\mathbf{p},\mathbf{a}_{pq})\right]$.

Using two inter-image homographies.
Given two views of two planar surfaces π_1 and π_2 and the related inter-image homographies between the two images, the line of intersection $\mathbf{l}$ between the two planes can be computed in both images in the following way: let $\mathtt{H}_1$ be the homography which maps points on π_1 on the first image into the corresponding plane in the second image, and $\mathtt{H}_2$ the homography which maps points on π_2 in the first image into the corresponding ones on the second image (see Fig. 6.17). Furthermore, let $\mathbf{l}$ be the intersection line in the first image and $\mathbf{l}'$ the corresponding line on the second image, then:

$$\mathbf{l}' = \mathtt{H}_1^{-\top}\mathbf{l}, \quad \mathbf{l}' = \mathtt{H}_2^{-\top}\mathbf{l} \tag{6.40}$$

from which [115]:

$$\tilde{\mathtt{H}}\mathbf{l} = \lambda \mathtt{I}\mathbf{l} \tag{6.41}$$

with $\tilde{\mathtt{H}} = \mathtt{H}_2^{\top}\mathtt{H}_1^{-\top}$. The 3×3 matrix $\tilde{\mathtt{H}}$ represents a homology, since it has a repeated eigenvalue (see section 3.1.3). The line $\mathbf{l}$ on the first image is the unique eigenvector corresponding to the different eigenvalue of the matrix $\tilde{\mathtt{H}}$. The intersection line $\mathbf{l}'$ on the second image may be computed simply by applying (6.40).

Using plane-plus-parallax.
In the previous approach, the inter-image homographies $\mathtt{H}_1$ and $\mathtt{H}_2$ induced by both planes are required. Employing the plane-plus-parallax approach, the problem of computing the line of intersection between two planes reduces to:

1. Choosing one plane as reference (e.g. π_1) and computing the inter-image homography induced by that plane between the two images (as usual from at least four correspondences)(e.g. the homography $\mathtt{H}_1$).
2. Defining three pairs of corresponding points on the second plane between the two images (they define the images of the tripoint plane).
3. Projecting the three points selected in the first image into the second by the homography $\mathtt{H}_1$ and projecting the three corresponding points selected in the second image into the first by $\mathtt{H}_1^{-1}$. Thus, in each image we have defined two shadow triangles. Those are in a Desargues configuration (see section 3.1.3), therefore corresponding sides interesect in three aligned points.

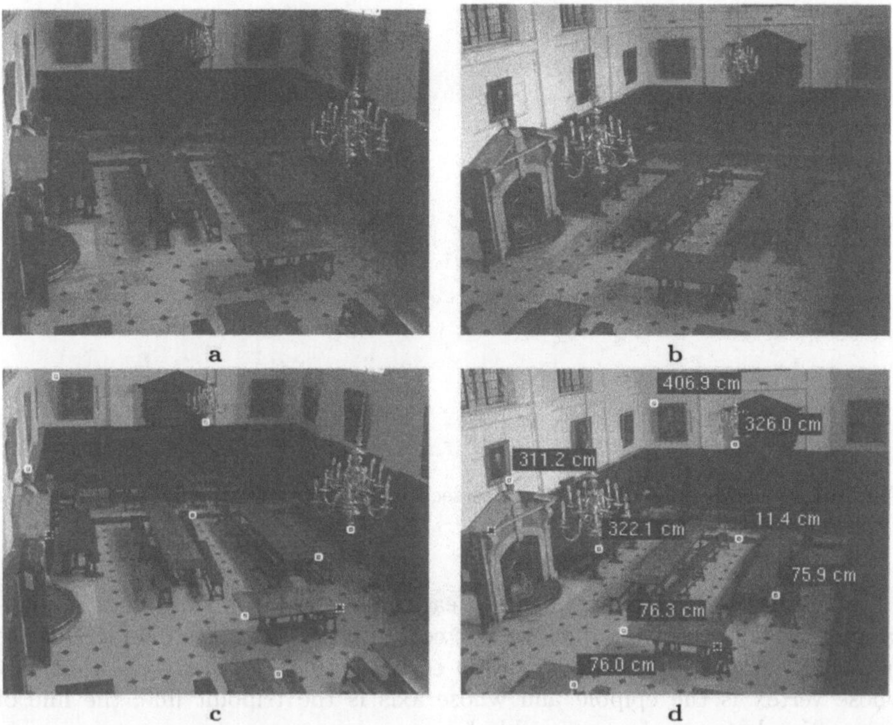

Fig. 6.16. Estimating the heights of points from two views of The Queen's College dining hall: **a** and **b** original left and right images; **c** and **d** the heights of the two reference points, shown black on white, were measured by hand to be 76 cm (table top) and 230 cm (fireplace). The heights of many other points have been computed by the parallax-based algorithm and the results shown in the image. Notice that all the top planes of the tables are about 76 cm from the floor.

4. The intersection lines **l** and **l**′ are the images of the tripoint line and are computed by fitting a line through the points of intersection of corresponding sides of the two shadow triangles in each image.

This approach has the advantage that the inter-image homography induced by the second plane does not need to be known, with only three point correspondences sufficient (affine mapping, see point 2 of the algorithm above). Likewise it is straightforward to prove that also two line correspondences on the second plane are sufficient.

Figure 6.18 shows an example. Figures 6.18**a** and **b** show two images of an indoor scene, taken from two different points of view. Figures 6.18**c** and **d** show the same images where we have marked the four points (corners of the poster) used to compute the inter-image homography induced by the left wall, the two shadow triangles (in Desargues configuration) and the computed intersection lines **l** and **l**′ in the two images.

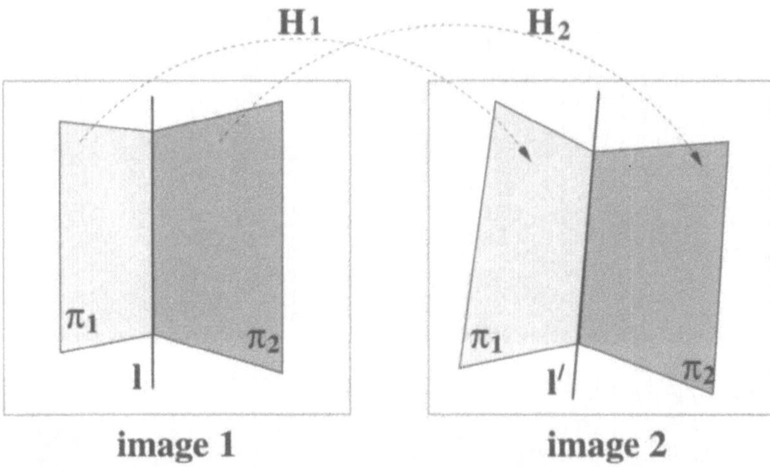

Fig. 6.17. Computing the line of intersection between two planar surfaces using planar homographies and two views.

Since the two shadow triangles in each image are in a Desargues configuration, in each image the lines joining corresponding vertices intersect in the epipole. Therefore the two triangles in each image are related by a homology whose vertex is the epipole and whose axis is the tripoint line, the line of intersection between the two world planes.

An alternative derivation of the intersection line is as follows: from six points (four coplanar), the fundamental matrix and the inter-image homography for one of the two planes are determined. From one more point off the plane the inter-image homography induced by the second plane is defined and therefore the first method can be applied.

6.4 Uncertainty analysis

As in section 5.3, here we analyze how input errors propagate through the computations in order to estimate the uncertainty of the final distance measurements. This is achieved by employing, once more, first order error analysis. The validity of the approximation is assessed in section 6.4.4 by Monte Carlo statistical tests.

The uncertainty analysis has been developed here for the case of distances computed directly from the images. However it can be easily extended to the distinguished plane case.

6.4.1 Uncertainty of camera distances

When computing distances of cameras from planes (6.35) uncertainty arises from the uncertain locations of the reference image pairs $\mathbf{p}$, $\mathbf{p}'$ and $\mathbf{q}$, $\mathbf{q}'$, the

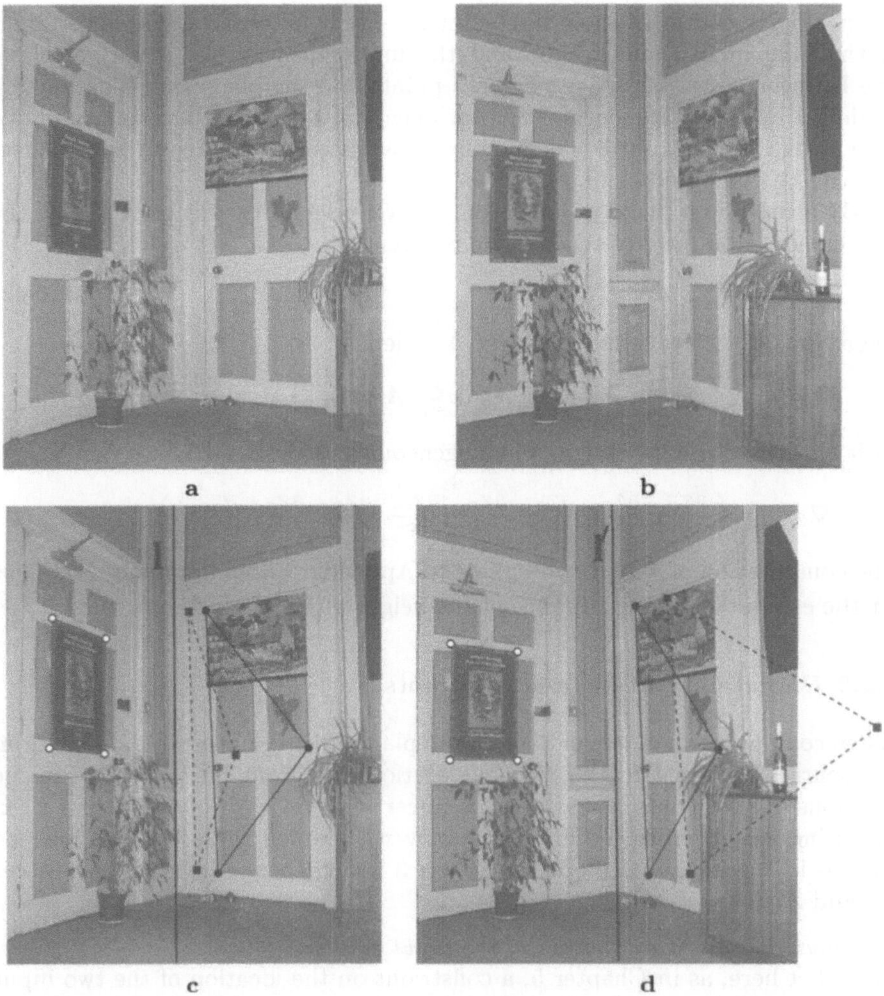

Fig. 6.18. Computing the line of intersection between two planar surfaces: **a** and **b** original left and right views of a corner of a room; **c** and **d** the four vertices of the poster on the left wall (marked with white circles) have been used to compute the inter-image homography induced by the left wall; the selected three corresponding points on the right wall (black circles joined by solid black lines) define the tripoint triangle; these points have been projected via the homography (or its inverse) into the other view (black squares joined by dashed black lines); the two shadow triangles are in a Desargues configuration. The lines of intersection **l** and **l**′ between the two walls have been computed by intersecting the corresponding sides of the two shadow triangles in each view and then joining up the intersection points.

uncertainty location of their three-dimensional counterparts, the uncertainty in the inter-image homography and the uncertainty in the plane vanishing line $\mathbf{l}$. Uncertainties in image reference points and the plane vanishing line are modelled by 3×3 homogeneous covariance matrices. The uncertainty of the inter-image homography $\mathtt{H}$ is defined by a 9×9 homogeneous covariance matrix and can be computed as in section 4.2.2.

By using a first order propagation, the variance of the distance Z_{O_1} of the first camera from the reference plane is given by:

$$\sigma^2_{Z_{O_1}} = \nabla_{Z_{O_1}} \Lambda \nabla^\top_{Z_{O_1}} \tag{6.42}$$

where the 26×26 covariance matrix Λ is defined as:

$$\Lambda = \text{diag}\left(\Lambda_{\mathbf{l}}, \Lambda_{\mathbf{h}}, \sigma^2_{Z_p}, \Lambda_{\mathbf{p}}, \Lambda_{\mathbf{p}'}, \sigma^2_{Z_q}, \Lambda_{\mathbf{q}}, \Lambda_{\mathbf{q}'},\right)$$

with all the covariance matrices homogeneous and the 1×26 Jacobian $\nabla_{Z_{O_1}}$:

$$\nabla_{Z_{O_1}} = \left(\frac{\partial Z_{O_1}}{\partial \mathbf{l}} \; \frac{\partial Z_{O_1}}{\partial \mathbf{h}} \; \frac{\partial Z_{O_1}}{\partial Z_p} \; \frac{\partial Z_{O_1}}{\partial \mathbf{p}} \; \frac{\partial Z_{O_1}}{\partial \mathbf{p}'} \; \frac{\partial Z_{O_1}}{\partial Z_q} \; \frac{\partial Z_{O_1}}{\partial \mathbf{q}} \; \frac{\partial Z_{O_1}}{\partial \mathbf{q}'} \right)$$

The computation of $\nabla_{Z_{O_1}}$ is described in Appendix F. Similar analysis applies for the estimation of the variance of the height of the second camera $\sigma^2_{Z_{O_2}}$.

6.4.2 Uncertainty of distance of points

When computing distances of points from planes (6.39), uncertainty arises from the sources analyzed in the previous section and from the uncertain image locations of the points $\mathbf{x}$ and $\mathbf{x}'$. The uncertainty in each of the points $\mathbf{x}$, $\mathbf{x}'$ (resulting largely from the finite accuracy with which these features may be located in the image) is modelled by 3×3 homogeneous covariance matrices $\Lambda_{\mathbf{x}}$ and $\Lambda_{\mathbf{x}'}$.

Maximum likelihood estimation of the input points.
Note that here, as in Chapter 5, a constraint on the location of the two input points $\mathbf{x}$ and $\mathbf{x}'$ exists. In this case the epipolar geometry constrains the two points to lie on each other's epipolar line. Therefore, as in section 5.3.2, a maximum likelihood estimate of the true points $\hat{\mathbf{x}}$ and $\hat{\mathbf{x}}'$ can be performed [57].

Not considering this intermediate step introduces a further approximation into the uncertainty propagation procedure. Nevertheless, the degradation of results caused by neglecting the maximum likelihood estimation has proven smaller than the approximation introduced by the first order truncation in the case that the two points $\mathbf{x}$ and $\mathbf{x}'$ are sufficiently close to the relative epipolar lines.

In order to keep the uncertainty analysis simple, then, the ML estimation of the true position of the input points has not been performed. The validity of the analysis is, as usual, assessed by Monte Carlo statistical tests.

Variance of distance.
By using a first order uncertainty propagation and assuming statistical independence in all the different sources of error, the variance of Z_x is given by:

$$\sigma^2_{Z_x} = \nabla_{Z_x} \Lambda \nabla^\top_{Z_x} \tag{6.43}$$

where the 32×32 covariance matrix Λ is:

$$\Lambda = \operatorname{diag}\left(\Lambda_{\mathbf{l}}, \Lambda_{\mathbf{h}}, \sigma^2_{Z_p}, \Lambda_{\mathbf{p}}, \Lambda_{\mathbf{p}'}, \sigma^2_{Z_q}, \Lambda_{\mathbf{q}}, \Lambda_{\mathbf{q}'}, \Lambda_{\mathbf{x}}, \Lambda_{\mathbf{x}'}\right)$$

with all the covariance matrices homogeneous and the 1×32 Jacobian ∇_{Z_x}:

$$\nabla_{Z_x} = \left(\frac{\partial Z_x}{\partial \mathbf{l}} \; \frac{\partial Z_x}{\partial \mathbf{h}} \; \frac{\partial Z_x}{\partial Z_p} \; \frac{\partial Z_x}{\partial \mathbf{p}} \; \frac{\partial Z_x}{\partial \mathbf{p}'} \; \frac{\partial Z_x}{\partial Z_q} \; \frac{\partial Z_x}{\partial \mathbf{q}} \; \frac{\partial Z_x}{\partial \mathbf{q}'} \; \frac{\partial Z_x}{\partial \mathbf{x}} \; \frac{\partial Z_x}{\partial \mathbf{x}'} \right) \tag{6.44}$$

Detailed description of the computation of ∇_{Z_x} is given in Appendix F.

6.4.3 Example

Figure 6.19 shows an example where heights of points and the associated uncertainties are computed for an indoor scene. Figures 6.19**a** and **b** are, respectively, left and right views of the interior of a chapel. We wish to measure the perpendicular distances of points of the lectern from the floor.

Two perpendicular heights have been measured by hand as references (marked in the images as white boxes). The inter-image homography between the two images has been estimated from four corresponding points on the floor tiling. Distances of several lectern points (marked as white circles) from the floor have been computed by using the described parallax method. A 3-standard deviation uncertainty ellipse has also been defined for each point, and then the 3-standard deviation uncertainty associated with the final height measurement has been estimated and displayed. Notice in Fig. 6.19**c** that the ground truth height always falls within the computed uncertainty range.

6.4.4 Validation of uncertainty analysis

This section validates the first order error analysis described above by estimating the uncertainty of the height of the beak of the eagle in the lectern in Fig. 6.19. It compares the results from the first order analytical method to the uncertainty derived from a Monte Carlo simulation, as described in Table 6.5. The diagrams in Fig. 6.20 illustrate the results.

In the simulations Gaussian noise is assumed on the image points of the two reference distances and on their measured world height. Uncertainty is assumed also on the plane vanishing line, on the inter-image homography and on the image points of the height to be computed.

The input point $\mathbf{x}$ on the left image is randomly distributed according to a 2D non-isotropic Gaussian about the mean location, with covariance matrix $\Lambda_{\mathbf{x}}$ (Fig. 6.20**a**). Similarly the corresponding point on the right image is randomly distributed according to a 2D non-isotropic Gaussian about the mean location $\mathbf{x}'$, with covariance $\Lambda_{\mathbf{x}'}$ (Fig. 6.20**b**).

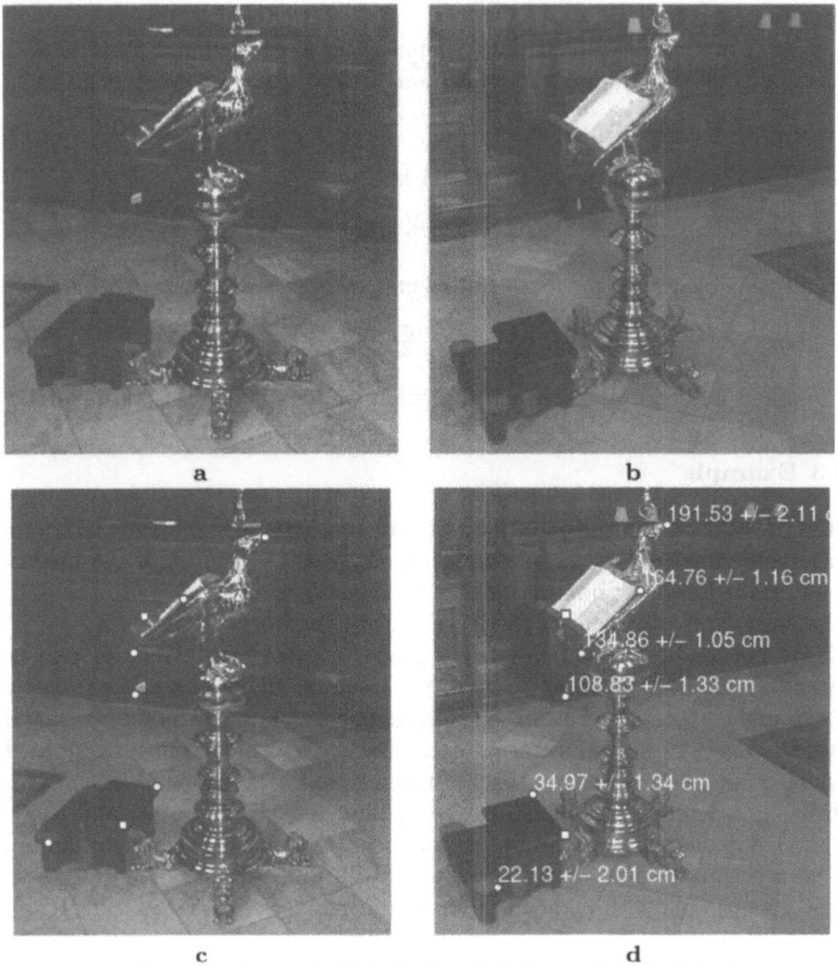

Measured	191.5	165.0	135.0
Computed	191.5 ± 2.1	164.7 ± 1.2	134.8 ± 1.0
Measured	109.1	35.0	22.9
Computed	108.8 ± 1.3	34.9 ± 1.3	22.1 ± 2.0

e

Fig. 6.19. Estimating heights and associated uncertainties from two views of the lectern in The Queen's College chapel: **a** and **b** original left and right images; **c** and **d** the heights of the reference points (white squares) were measured by hand (± 0.5 cm) to be 150 cm (top edge of lectern), and 35 cm (height of foot stool). The heights of many other points have been computed and superimposed to the images. In **e**, all height measurements are given in centimetres. The error between the heights computed using the image parallax and the ones measured manually is always less than one centimetre. The 3-standard deviation uncertainties of the computed heights are also shown. The ground truth falls always within the computed uncertainty range.

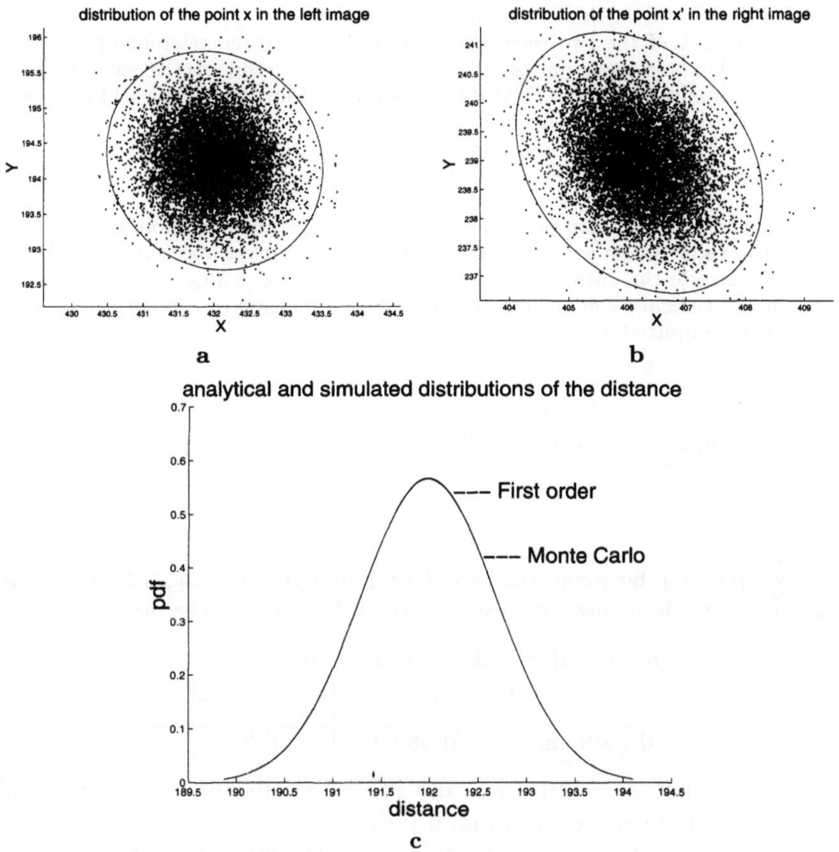

Fig. 6.20. Monte Carlo statistical test of the example in Fig. 6.19. **a** Distribution of the input point $\mathbf{x}$ on the left image and the corresponding 3-standard deviation ellipse. **b** Distribution of the input point $\mathbf{x}'$ on the right image and the corresponding 3-standard deviation ellipse. Figures **a** and **b** are drawn at the same scale. **c** The analytical and simulated distributions of the computed distance Z. The two curves overlap almost perfectly.

The two covariance matrices for the pair of input points are respectively (values in pixel^2):

$$\Lambda_{\mathbf{x}} = \begin{pmatrix} 0.255 & -0.024 \\ -0.024 & 0.266 \end{pmatrix} \quad \Lambda_{\mathbf{x}'} = \begin{pmatrix} 0.484 & -0.148 \\ -0.148 & 0.566 \end{pmatrix}$$

Suitable values for the covariances of the two references, and the vanishing line, have been used. The covariance of the inter-image homography has been computed as in section 4.2.2. The simulation was run with $N = 10\,000$ samples.

Analytical and simulated distributions of Z_x are plotted in Fig. 6.20**c**; the two curves are almost overlapping. Any slight difference is due to the assumptions of statistical independence and first order truncation introduced by the error analysis.

- For j=1 to N (with N = number of samples)
  - For each of the two references: given the measured reference points on the left and right image, generate a random point on the left image, a random point on the right image and a random reference world distance according to the associated covariances.
  - Generate a random plane vanishing line according to its covariance $\boldsymbol{\Lambda}_{\mathbf{l}}$.
  - Generate random values for the inter-image homography $\tilde{\mathtt{H}}$ according to the computed 9×9 covariance matrix $\boldsymbol{\Lambda}_{\mathbf{h}}$.
  - Generate a random point $\mathbf{x}_j$ on the left image and a random point $\mathbf{x}'_j$ on the right image according to their respective covariances $\boldsymbol{\Lambda}_{\mathbf{x}}$ and $\boldsymbol{\Lambda}_{\mathbf{x}'}$.
  - Compute the current distance Z_{x_j} by applying (6.35).
- The statistical standard deviation of the population of simulated Z_{x_j} values is computed as:

$$\sigma'^2_Z = \frac{\sum_{j=1}^{N}(Z_j - \overline{Z})^2}{N-1}$$

and compared to the analytical one (6.43).

Table 6.5. Monte Carlo test.

A comparison between statistical and analytical standard deviations is shown in the table below with the corresponding relative error:

First Order σ_{Z_x}	Monte Carlo σ'_{Z_x}	relative error $\frac{\lvert\sigma_{Z_x} - \sigma'_{Z_x}\rvert}{\sigma'_{Z_x}}$
0.7039 cm	0.7056 cm	0.24%

The measurement $Z = 191.8$ cm and the associated first order uncertainty $3 * \sigma_{Z_x} = 2.11$ cm is shown in Fig. 6.19**d**.

Similar tests on several other examples consistently yielded the same results. This shows that in the plane-plus-parallax approach also, for typical imaging situations the first order approximation is a valid one.

As usual, some care must be exercised since the relative error between statistical and analytical output standard deviations increases with the input uncertainty. For large covariances, the assumption of linearity becomes less well founded.

7. Gallery of examples

This chapter shows examples of three-dimensional reconstructions obtained by employing the techniques described in Chapter 5. The algorithms have been applied to both photographs of real scenes and perspectively correct paintings. From the computed models, convincing animations and fly-throughs have also been created.[1]

7.1 Reconstruction from photographs

Images of the three-dimensional model of a garden shed reconstructed from a single photograph appear in Fig. 7.1. A colour version of this figure is shown in colour plate 1. The reconstruction algorithm employed is described in section 5.5.

The height and position of the person are computed correctly but, since it is not possible to recover volume from one image alone, the person is represented as a flat silhouette. The outline of the silhouette was extracted by fitting a spline curve to the contour of the image.

In the future, we may attempt to reconstruct the volume of people from single images by making use of: shading, silhouette and prior assumptions about the shape of the human body. This would help produce more realistic three-dimensional environments.

7.2 Reconstruction from paintings

Linear perspective can be modelled mathematically by projective geometry. If a painting conforms to the rules of linear perspective (see section 1.4.3) then it behaves geometrically as a perspective image and it can be treated as a photograph. The metrology techniques previously described can thus be applied to compute the structure of the scene. Since no Euclidean scene measurements are known (usually the scenes depicted do not exist), the models are obtained

[1] The reconstructed three-dimensional environments can be "virtually explored" in the VR-CUBE at KTH, Stockholm. VRML three-dimensional models and MPEG animations can also be viewed at the following web pages:

`http://www.robots.ox.ac.uk/~vgg/projects/SingleView/`

`http://www.research.microsoft.com/Users/antcrim/`

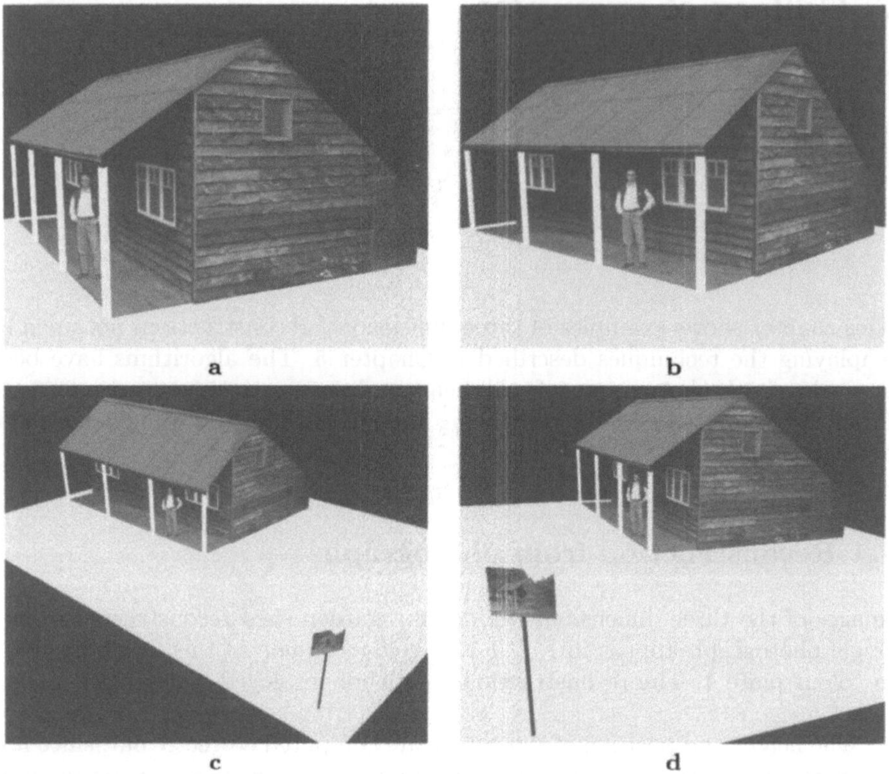

Fig. 7.1. Three-dimensional reconstruction of a garden shed. **a** and **b** two views of the reconstructed three-dimensional model of the shed in Fig. 5.23**a**. See also colour plate 1. The person is represented as a flat silhouette since it is not possible to recover its volume from one image. The silhouette has been cut out manually from the original image and positioned in the correct place. **c** and **d** are two more views of the three-dimensional model showing the computed camera position.

up to a similarity transformation. This level of reconstruction is precisely that required for a graphical three-dimensional model where the absolute pose and scale are not necessary for visualization.

Before reconstructing a painting though, it is necessary to check the correctness of its perspective. This has been assessed here simply by computing the location of vanishing points and vanishing lines. Perspective distortion in repeated patterns (e.g. the rate of diminution of the coffers on the curved vault in Fig. 7.2**a**, or the tiles on the floor in Fig. 7.4**a**) has been checked by using cross-ratios.

Fig. 7.2. Three-dimensional reconstruction of *Trinità*: **a** *Trinità con la Vergine e San Giovanni* (approx. 1426), by Masaccio (Tommaso di Ser Giovanni Guidi, 1401–1428), Museo di Santa Maria Novella, Firenze, Italia. Courtesy of the Musei Comunali di Firenze. **b** A view of the reconstructed three-dimensional model of the chapel from the right. Notice the hemi-cylindrical vault. **c** View of the model from the left. **d** View of the model from below. **e** A more extreme view from the left side. The left wall has been removed to show the inside structure of the chapel. See also colour plate 2.

7.2.1 *Trinità*, Masaccio

Trinità (in Fig. 7.2**a**) is a fresco. It is, arguably, the first perspective image in history; the first example of the application of Filippo Brunelleschi's rules of perspective (cf. colour plates 2 and 3).

A three-dimensional model of the chapel represented in the fresco has been computed here by applying the algorithm described in Chapter 5. Different views of the model are shown in Fig. 7.2**b**–**e**. Notice the hemi-cylindrical vault. The parts of the vault which are occluded by the capitals and by the head of God have been left blank. Their texture can be filled in by making use of the regularity of the pattern.

Ambiguity in recovering depth.
Since one image alone is used and no scene metric information is known (the chapel is not real), an ambiguity arises in the reconstruction: it is not possible to recover the depth of the chapel without making some assumptions about the geometry of the scene. Two equally valid assumptions can be made: either the coffers on the vault of the chapel are square (this is the assumption used in the model in Fig. 7.2) or the floor is square.

Both models have been computed (the two reconstructions are mathematically related by a simple affine transformation, a scaling in the direction orthogonal to the plane of the fresco). Images of the floor of the chapel and of the pattern on the cylindrical vault are shown in Fig. 7.3 for both cases.

This demonstrates that the two assumptions cannot coexist in the same reconstruction, even though the ambiguity is not apparent to the naive observer. Speculations about which of the two configurations is the one that most likely the artist wanted to represent are left to the reader.

7.2.2 *Flagellazione*, Piero della Francesca

Flagellazione (in Fig. 7.4**a**, and also in colour plates 4 and 5) is one of the most studied paintings from the Italian Renaissance period. Mainly because of the "obsessive" correctness of its perspective [67]. In the past, the painting has attracted the interest of several artists and art historians who, in different ways, have been studying its carefully represented geometry.

Since this painting conforms faithfully to the rules of linear perspective the depicted scene can correctly be reconstructed in three dimensions by applying the algorithms described in Chapter 5. Images of the computed model, taken from different viewpoints, are shown in Fig. 7.4**b**–**d**.

7.2.3 *St Jerome in His Study*, H. Steenwick

St Jerome in His Study (in Fig. 7.5**a**, and also in colour plates 6 and 7) is an oil painting by the Dutch artist H. Steenwick (1580–1649). It is particularly interesting for two reasons: the correctness of the perspective (when this masterpiece was realized linear perspective was a well-established theory) and the amazing management of light and shading. The sunlight streaming through the window gives the painting a beautiful photo-realistic touch.

Numerous parallel lines and planar surfaces have been detected and employed to produce the reconstruction results illustrated in Fig. 7.5**b**–**f**.

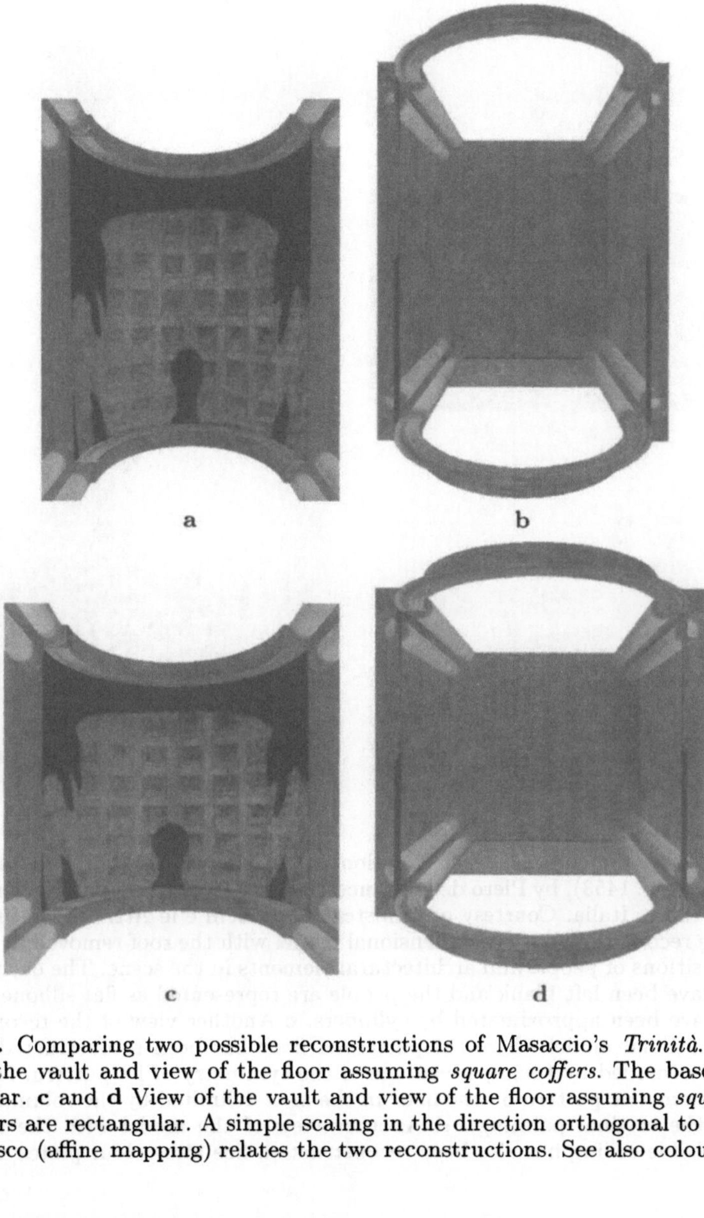

Fig. 7.3. Comparing two possible reconstructions of Masaccio's *Trinità*. **a** and **b** View of the vault and view of the floor assuming *square coffers*. The base plane is rectangular. **c** and **d** View of the vault and view of the floor assuming *square floor*. The coffers are rectangular. A simple scaling in the direction orthogonal to the plane of the fresco (affine mapping) relates the two reconstructions. See also colour plate 3.

Detecting inconsistencies in the perspective rendering.
Figure 7.6**a** shows a fronto-parallel view of the big window on the left wall of the scene. This image has been obtained, as usual, by rectifying the original texture in the painting by inferring affine and metric information (parallel lines and right angles) on the wall plane.

Notice that while parallelism and angles have been recovered correctly (look at the window pattern) an unexpected, asymmetric curvature of the top arch

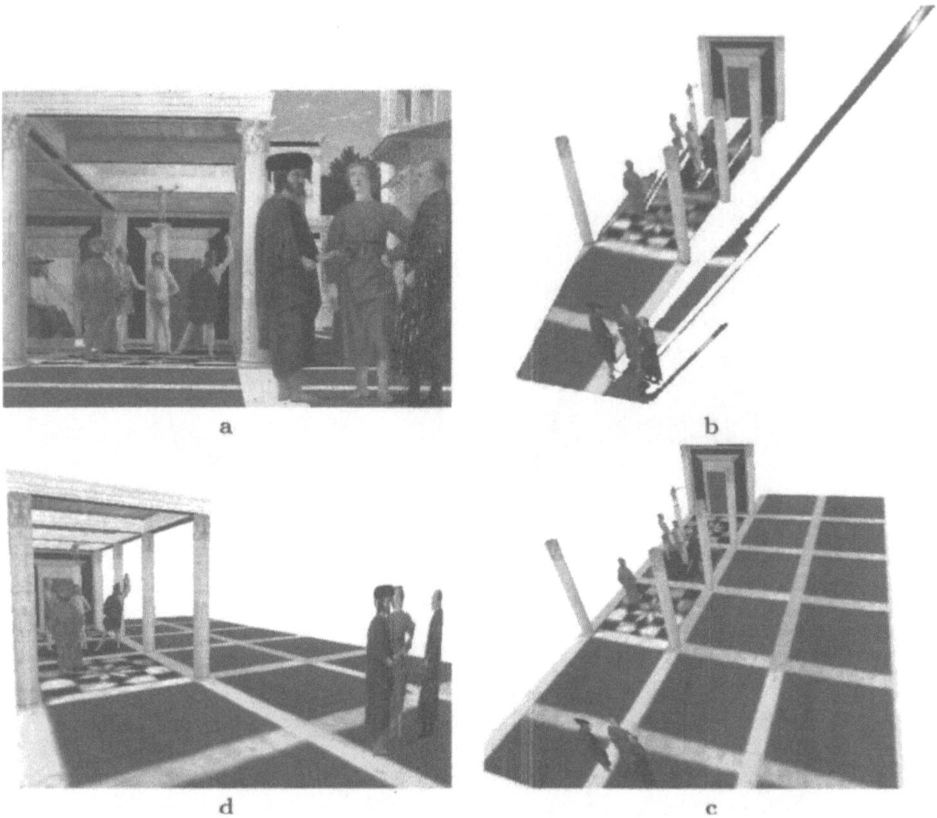

Fig. 7.4. Three-dimensional reconstruction of *Flagellazione*. **a** The painting *Flagellazione* (approx. 1453), by Piero della Francesca (1416–1492), Galleria Nazionale delle Marche, Urbino, Italia. Courtesy of Ministero per i Beni e le Attività Culturali. **b** A view of the reconstructed three-dimensional model with the roof removed to show the relative positions of people and architectural elements in the scene. The occlusions on the floor have been left blank and the people are represented as flat silhouettes. The columns have been approximated by cylinders. **c** Another view of the reconstructed three-dimensional model where the patterned floor has been reconstructed in areas where it is occluded by taking advantage of the symmetry of its pattern. Notice the repeated geometric pattern on the floor in the area delimited by the columns (barely visible in the original painting). **d** Another view of the model with the roof. The partially visible ceiling has also been reconstructed. See also colour plates 4 and 5.

can be detected. That is due to the fact that the artist has painted a fairly complicated curve on a very slanted view of the window (cf. the original painting in Fig. 7.5**a**). Large uncertainty characterizes the localization of points in such a situation (cf. examples in section 4.3.2). The error is less evident in views taken from locations closer to the original viewpoint (Fig. 7.6**b**).

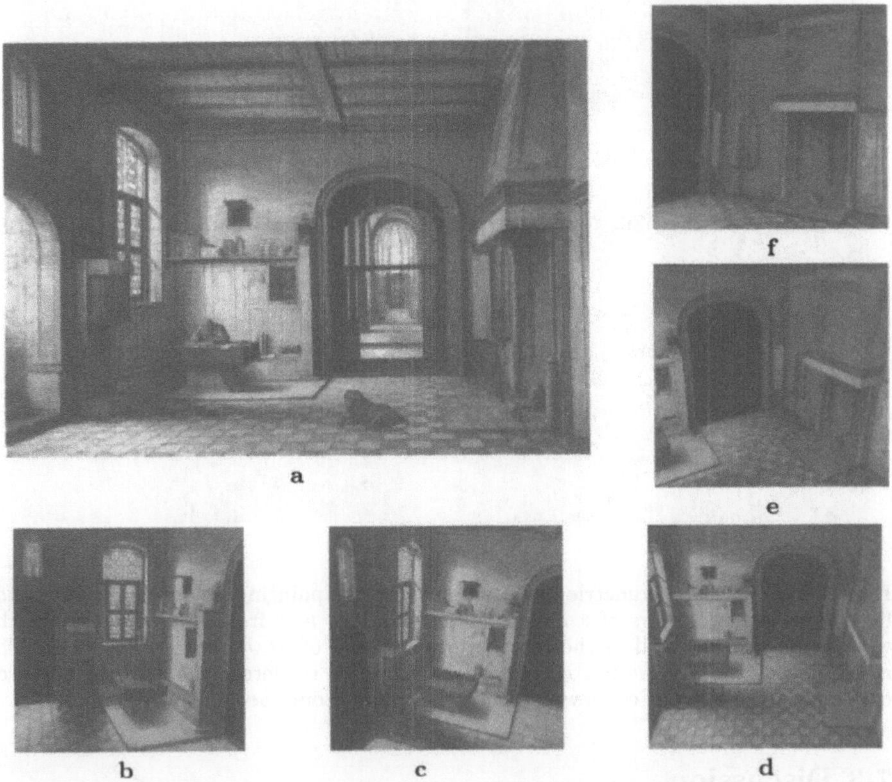

Fig. 7.5. Three-dimensional reconstruction of *St Jerome in his Study*: **a** *St Jerome in his Study* (1630), by Hendrick V. Steenwick (1580–1649), private collection (Joseph R. Ritman Collection), Amsterdam, The Netherlands. **b-f** Different views of the reconstructed three-dimensional model. **b** The left wall; the the partially seen windows have been reconstructed by making use of the symmetrical pattern in their textures. **c** The corner where St Jerome is working. **d** View of the right-hand side of the study. **e** View of the fireplace from above. **f** Another view of the fireplace. See also colour plate 6.

7.2.4 *The Music Lesson*, Vermeer

The Music Lesson (in Fig. 7.7**a** and colour plate 8) is another example of fine Dutch perspective. Here too we find a perfect geometric construction of the depicted indoor scene (probably real), together with a very skilled use of light and an in-depth knowledge of its effects on the overall atmosphere of the painting [92].

Notice that some of the fairly complex objects (e.g. the cello, the chair) represented in the painting have not been reconstructed in the simple final model.

a b

Fig. 7.6. Detecting geometric inconsistencies in the painting *St Jerome in his Study* by comparing two views of a reconstructed window. **a** A fronto-parallel view of the window on the left wall of the reconstructed model of *St Jerome in his Study*. The right part of the arch is not correct. **b** It is harder to detect the error in the arch curvature from a point of view closer to the original one. See also colour plate 7.

7.3 Discussion

At this point I would like to invite the reader to reflect on a particular aspect of the relationship between science and art.

Some computer vision researchers describe their field as a combination of *maths*, *science* and *engineering* [82]. But is that all? Is there any room left for *art*? Can art help us understand vision or vice-versa?

For instance, when looking at photographs of a scene, visual cues such as converging straight lines, shading effects, texture, shadows and specularities are processed by our brain to retrieve consistent information about the surrounding three-dimensional world.

The same cues are employed by artists in their paintings. However, since these works of art are not generated by an automatic imaging process (such as a camera) but by the skilled hands of a painter, they are prone to personal interpretation and inaccuracies. These visual signals might therefore not be consistent with each other (e.g. images of parallel lines may not intersect in exactly the same point). However, as a large amount of physiological research shows, our brain is capable of neglecting conflicting perspective cues; therefore,

Fig. 7.7. Three-dimensional reconstruction of another Dutch painting *The Music Lesson*. **a** *A lady at the virginals with a gentleman (the music lesson)* (1662-65), by Johannes Vermeer (1632-1675), The Royal Collection © 2000, Her Majesty Queen Elizabeth II, London. **b-f** Different views of the reconstructed three-dimensional model. Some objects in the painting do not appear in our reconstruction (e.g. the cello and the chair). See also colour plate 8 and note the beautiful lighting that dominates the painting. Highly skilled perspectival techniques are demonstrated by the artist in the rendering of the floor pattern.

slightly inaccurate perspective paintings may still convey the desired three-dimensional illusion. [2]

[2] see for instance the *Arnolfini Portrait* (1434, by Jan van Eyck (active from 1422–1441), National Gallery, London), or the *Martyrdom of Savonarola* (by anonymous, San Marco Museum, Firenze, Italy)

A number of questions arise: Which perceptual cues are more important to the three-dimensional illusion? To what extent do humans forgive wrong cues? Since paintings and single photographs are capable of conveying an illusion of the three-dimensional space, how necessary is stereo vision? These points may lead the way to further physiological speculations.

On the other hand projective geometry can provide artists and art historians with fast and powerful mathematical tools (e.g. planar projections, cross-ratio) to answer questions like: How do I construct the geometry of the painting such that the perspective looks correct when viewed from such a viewpoint? How correct is the perspective (e.g. vanishing points, perspective distortion) in Luca Signorelli's *Circoncisione*? [3] What does the pattern on the floor in Domenico Veneziano's *Pala di Santa Lucia*[4] look like? Is the height of the foreground figures consistent with the height of the background figures in Raffaello's *Sposalizio*? [5] And how does it compare to the *Sposalizio* by Perugino? [6]

In my opinion computer vision, art and art history are well-distinguished fields, each with its own aims and motivations. Nevertheless, each might learn from and be enriched by the others. Furthermore, the tools developed in one area may be useful in another.

[3] *Circoncisione*, by Luca Signorelli (1445-1461), National Gallery, London

[4] *Santa Lucia dei Magnoli Altarpiece*, 1444, by Domenico Veneziano (1400-1461), Uffizi, Firenze, Italy

[5] *Sposalizio*, 1504, by Raffaello Sanzio (1483-1520), Brera Museum, Milano, Italy

[6] *Il Matrimonio della Vergine*, 1500-1504, by Pietro Perugino (1450-1523), Musee des Beaux Arts, Caen, France

Colour plate 1

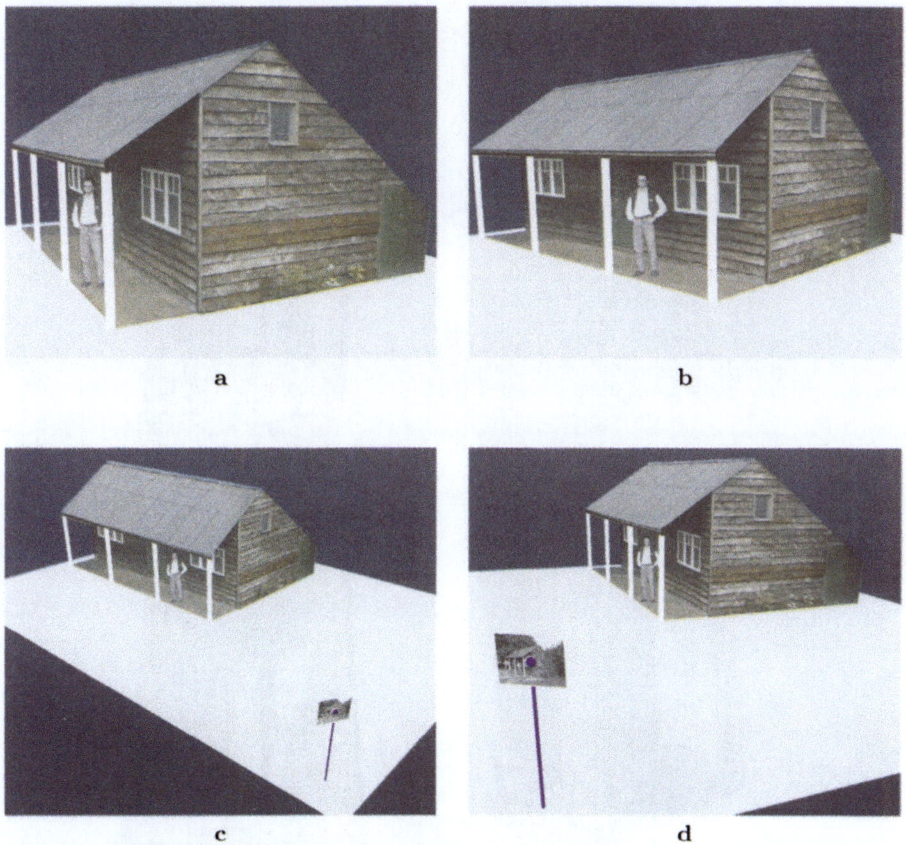

Three-dimensional reconstruction from a single, uncalibrated photograph. **a** and **b** Two views of the reconstructed three-dimensional model of the shed in Fig. 5.23**a**. The reconstruction algorithm is described in Chapters 5 and 7. Note that the person is represented as a flat silhouette. **c** and **d** Two more views of the three-dimensional model showing the computed, camera position.

Colour plate 2

Three-dimensional reconstruction of Masaccio's *Trinità*: **a** The original painting: *Trinità con la Vergine e San Giovanni* (approx. 1426), by Masaccio (Tommaso di Ser Giovanni Guidi, 1401–1428), Museo di Santa Maria Novella, Firenze, Italia. Courtesy of the Musei Comunali di Firenze. The single-view reconstruction algorithm is described in Chapter 5 and 7 and results are also shown in Fig. 7.2. **b-e** Different views of the reconstructed three-dimensional model of the chapel in the Florentine fresco.

Colour plate 3

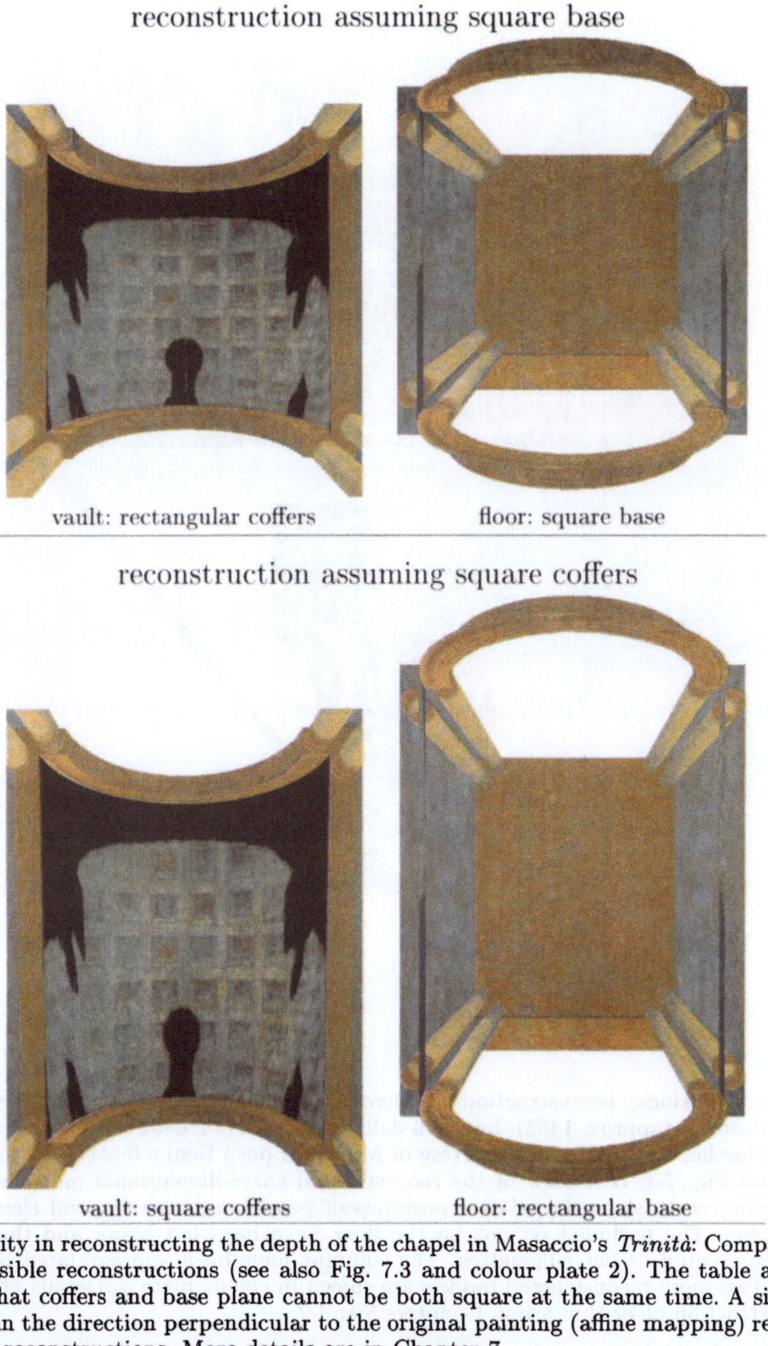

Ambiguity in reconstructing the depth of the chapel in Masaccio's *Trinità*: Comparing two possible reconstructions (see also Fig. 7.3 and colour plate 2). The table above shows that coffers and base plane cannot be both square at the same time. A simple scaling in the direction perpendicular to the original painting (affine mapping) relates the two reconstructions. More details are in Chapter 7.

Colour plate 4

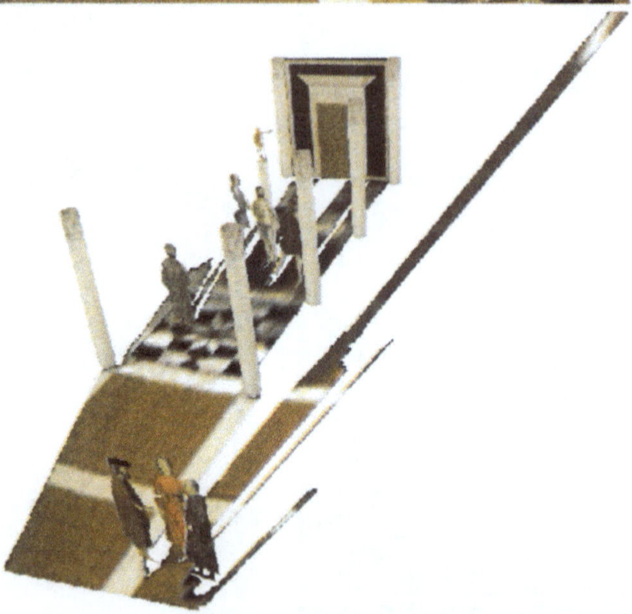

b

Three-dimensional reconstruction of Piero's *Flagellazione*: **a** the original painting *Flagellazione* (approx. 1453), by Piero della Francesca (1416–92), Galleria Nazionale delle Marche, Urbino, Italia. Courtesy of Ministero per i Beni e le Attività Culturali; See also Fig. 7.4. **b** A view of the reconstructed three-dimensional model with the roof removed to show the relative positions of people and architectural elements in the scene. The occluded regions on the floor have been left blank and the people are represented as flat silhouettes. The columns have been approximated by cylinders. Two more reconstructed models are shown in colour plate 5. Details about the reconstruction algorithm are in Chapters 5 and 7.

Colour plate 5

c

d

Three-dimensional reconstruction of Piero's *Flagellazione*: Two more views of the reconstructed three-dimensional model (see also colour plate 4 and Fig. 7.4). **c** The patterned floor has been reconstructed in areas where it is occluded by taking advantage of the symmetry of its pattern. Notice that the geometric pattern on the floor appears at both the back and front of the region enclosed by the columns (this pattern is barely visible in the original painting). **d** The partially visible ceiling has also been reconstructed.

Colour plate 6

Three-dimensional reconstruction of a Dutch painting. **(above)** The original painting *St Jerome in his Study* (1630), by Hendrick V. Steenwick (1580–1649), private collection (Joseph R. Ritman Collection), Amsterdam, The Netherlands. **(right)** The column on the right shows different views of the reconstructed three-dimensional model. See chapter 5 and 7 for more details on the algorithm, results are also shown in Fig. 7.5.

Colour plate 7

Detecting geometric inconsistencies in the painting *St Jerome in his Study* by comparing two views of a reconstructed window. **(upper)** a fronto-parallel view of the window on the left wall of the reconstructed model. The right part of the arch is not correct, it is not symmetric (highlighted on the right hand image). **(lower)** It is harder to notice the error in the arch curvature from a viewpoint closer to the original one. See also Fig. 7.6.

Colour plate 8

Three-dimensional reconstruction of a more recent Dutch painting. **(above)** The original painting *A lady at the virginals with a gentleman (the music lesson)* (1662–65), by Johannes Vermeer (1632–1675), The Royal Collection © 2000, Her Majesty Queen Elizabeth II, London. **(right)** The column on the right shows different views of the reconstructed three-dimensional model. The reconstruction algorithm is described in Chapter 5. See also Fig. 7.7.

8. Conclusion

8.1 Summary

This book has developed new, flexible and accurate algorithms for the use of a camera as a three-dimensional measuring device.

A solid theoretical framework employing techniques drawn from projective geometry has been developed to compute measurements from single and multiple uncalibrated views.

The process of measuring has been treated as a true engineering task, therefore particular attention has been paid to predicting the uncertainty of the final measurements arising from the uncertainty of the input data and of the measuring process.

The theory in this book has been developed in steps: from simpler metrology on planar surfaces to more complicated three-dimensional measurements and from single to multiple images.

Metrology on planes.
Earlier in this book we showed how to compute measurements on planar surfaces (Chapter 4). Distances between points and parallel lines, image rectification and image mosaicing have been discussed in detail.

The effectiveness of such techniques has been demonstrated with an application capable of computing measurements from single images of indoor and outdoor scenes, as well as for rectifying projectively distorted views and mosaicing together several images of a planar structure.

Planar projective mappings, such as homographies and homologies, have proven extremely useful, simple and powerful mathematical tools and they have been employed throughout the book.

Single-view metrology.
We have also explored how three-dimensional affine measurements can be retrieved from single views (Chapter 5). The key has been the novel parametrization of the three-dimensional space as a collection of pencils of parallel planes. Three-dimensional geometry has been successfully described in terms of the relationships between homographies, homologies and 3D-2D projective mappings.

This approach has made possible the reconstruction of the three-dimensional space in a stratified fashion. The available scene information may be used to upgrade the computed structure from affine to Euclidean.

This work has been applied successfully to a number of different areas: scientific, artistic and commercial. Software applications employing some of the presented algorithms have been developed and they are currently used for forensic purposes.

Metrology from planar parallax.
The use of a multi-view approach in cases where single-view techniques are less adequate has been addressed.

We have investigated the use of a plane-plus-parallax approach to make optimal use of the redundancies between different views of a scene. This has led to the discovery of elegant duality relationships between configurations of points and cameras, as described in Chapter 6.

Furthermore, we have demonstrated that the fundamental geometric constraints (epipolar geometry and trifocal tensor) are captured by a planar homology relating the images of points across views. Consequently, a number of previously diverse results related to planar parallax, duality and planar homologies have been unified.

Formulae for determining distances of cameras and points from a world plane have been presented, with the proper uncertainty analysis, and have been tested on real images.

8.2 Discussion

Advantages and disadvantages of the proposed metrology algorithms are discussed in this section.

An accurate and reliable measuring device.
The internal parameters of the camera are quite unstable and sensitive to variation in temperature or mechanical shock. Metrology algorithms which rely on the calibration of the camera are therefore strongly affected by such instability.

The algorithms proposed in this book do not make use of internal calibration, thus leading to greater robustness and reliability. Since no knowledge about the cameras is needed, the algortihms can be applied to existing, archived images.

The accuracy of the measuring device described in this book has been modelled mathematically, thus providing a tool to assess the precision of the output measurements, given uncertain input data and transformations. Moreover, the uncertainty analysis provides a powerful tool to increase the accuracy of the measurements. In fact, by observing the behaviour of the predicted accuracy, the ideal location of camera and references can be chosen such that the output uncertainty is minimized. This makes the device particularly effective in typical engineering or architectural applications such as modelling buildings and interior design (see section 4.3).

As expected, the quality of the results increases with the resolution of the images. Nevertheless, a 1024×768 resolution has proven sufficient for the applications considered.

Robustness for noisy images.
The images that the metrology device is required to deal with may contain a large amount of random noise. This happens quite often in forensic data, where the images are recorded on old tapes.

Nevertheless, useful geometric cues (like edges) can still be detected even for large amount of noise. The precautions during the process of detecting lines and vanishing points (Canny edge detection, straight-line fitting, edge merging and maximum likelihood estimates of intersection points, cf. sec. 3.3) provide the proposed metrology algorithms with an increased robustness and insensitivity to noise.

Flexibility.
Since no internal camera calibration is required at any time this device can be applied to a wide range of input images: old footage, archived photographs, postcards and even paintings. Effectively the algorithms can deal with any perspective image, no matter how that is obtained or where it comes from.

This is demonstrated, in particular, by the interest shown by researchers in fields like art history towards the capabilities offered by such techniques. Artists and art historians are becoming increasingly aware of the potential of modern technology. Some of the most fascinating applications, such as taking measurements on a painted scene (real or not), reconstructing it as a virtual reality model and creating animations and fly-throughs have been presented in this book.

As mentioned above, these strongly geometrical techniques can only be applied to projective images. Therefore they are particularly useful when applied to paintings from the Italian Renaissance whose masters invented Linear Perspective [1]. Numerous perspective paintings are also found amongst the Dutch masterpieces of the seventeenth century and in some contemporary works of art.

Apart from the obvious applications in art history and computer graphics, analyzing paintings can be a useful aid to the understanding of vision and the perception of the world around us.

Radial distortion correction.
When a camera becomes part of a measuring device, the accuracy and sensitivity of the camera itself has to be investigated. In particular, high-quality lenses and high-resolution CCD arrays may be required in order to get the most precise information from the viewed scene.

Quite often, images taken with wide-angle-lens cameras (often pictures of indoor scenes) are corrupted by lens-generated distortions such as radial distortion. These images, therefore, are not perspective. The unthinking application of calibrated or uncalibrated metrology techniques would therefore yield incorrect results. This problem has been overcome, in this book, by employing a simple and robust preprocessing stage, where a radial correction algorithm (section 3.2) warps the image into a perspective one.

This correction algorithm, though, is of no use when the camera lenses show more complicated aberrations. Most modern digital still or video cameras do not suffer, however, this problem and therefore the correction step is unnecessary.

Automation.
The proposed metrology algorithms must be supervised. In fact, features of interest in images (e.g. edges and corners) are detected automatically and localized with great accuracy, but interactivity is necessary to make the right inference from the images: selecting a set of parallel lines or a set of planar points (orthogonality is not necessary); identifying reference distances and entering their value; and selecting the end points of the length to be measured.

While complete automation is desirable in the long term, and indeed some of the stages which are currently supervised could be automated (e.g. detecting concurrent lines and vanishing points [97]), this has not been a major issue in this book.

8.3 Future work

The described measurement and structure recovery algorithms work particularly well for man-made environments. It is more complicated to apply those techniques to images of scenes where fewer regular structures and well defined geometric cues (e.g. parallel lines and planes) are present. In such cases different sources of geometric information must be used.

Use of shadows, reflections, symmetries and repetitions in single views.
As mentioned in section 5.6, shadows, reflections, symmetries and repetitions constitute an important source of pure geometric information (see Fig. 8.1). These cues provide restrictions analogous to a second viewpoint. It has been proven that projective or affine structure can be computed from single images with a plane of symmetry (e.g. a mirror, a lake) [60, 104]. Curved surfaces and non-planar objects can be reconstructed from single views by making good use of those geometric cues.

Shadows, reflections, symmetries of planes and repetitions can be described by homologies. One of the future goals of this work is, then, to build a systematic interpretation of the single-view approach which makes further use of the generality of the homology mapping.

A possible application is the reconstruction of complicated vault structures in churches from postcards, drawings and paintings [91]. For instance, in Fig. 8.2**a** the symmetry and repetition of the three-dimensional pattern in the vaults carry useful information about their shape and geometry.[1]

Being able to recover the shape of non-planar and generally curved objects may lead to an increase in the amount of detail present in reconstructed scenes.

[1] Note that the arches in Fig. 7.2 and Fig. 7.5 have been reconstructed with ease because they lie on fronto-parallel planes. In Fig. 7.2, the arch is simply a semicircle.

Fig. 8.1. Reconstructing curved surfaces from single views using reflections: a view of the bridge in the University Parks in Oxford. The reflection of the bridge on the water of the river provides information about the geometry of the bridge itself.

Analyzing natural scenes.
Currently we are trying to address the problem of dealing with "natural" scenes, where no strong geometric cues or straightforward geometric relationships (e.g. homologies) can be identified. This will probably require a statistical interpretation of the texture and shading of surfaces.

Less strictly geometry-based approaches may be applied in interpreting images of countryscapes and paintings. These images do not present many reliable geometric fixtures, but other less well-defined cues may convey the perception of depth, distance and shape of surfaces. Shape can be perceived, for instance, from the texture of surfaces [27, 49, 66] (see Fig. 8.2**b**, and also the famous "water-lilies" paintings by Claude Monet), and depth and distance from the variation of the intensity and saturation of the light. [2] This "atmospheric cue" might prove useful, not only to measure relative distances between objects (like buildings [90]) but, if reference distances are known, to measure the humidity or pollution of the air itself.

Uncertainty analysis with bootstrap methods.
On the error analysis front we are investigating alternative methods for predicting the uncertainty of the final measurements which avoid the sometimes tedious computation of Jacobian matrices required by first order error propagation techniques.

One possibility is to use numerical algorithms. *Bootstrap* methods [83] are currently under investigation. These techniques are very similar to the Monte Carlo approach, the main difference being that no knowledge about the statistical distribution of the input data is required (in this book, Gaussian noise

[2] *"There is another kind of perspective which I call aerial, because by the difference in the atmosphere one is able to distinguish the various distances of different buildings, which appear based on a single line." Leonardo da Vinci (1452–1519).* (cf. Leonardo's painting *The Virgin of the Rocks*, 1508, National Gallery, London.)

a b

Fig. 8.2. Reconstructing curved surfaces from single views using symmetry and texture: **a** a postcard showing the interior of the Bodleian Library in Oxford; reconstructing the beautiful and fairly complicated vaults is a challenge; **b** a photograph of the water-lilies pond in the University Parks in Oxford.

has always been assumed in the input data). Bootstrap algorithms, however, like the Monte Carlo ones, are based on an iterative refinement of the solution and therefore can be very slow.

"Questa conclusione, benché trovata da povera gente, c'è parsa cosí giusta, che abbiam pensato di metterla qui, come il sugo di tutta la storia.
La quale, se non v'è dispiaciuta affatto, vogliatene bene a chi l'ha scritta, e anche un pochino a chi l'ha raccomodata. Ma se in vece fossimo riusciti ad annoiarvi, credete che non s'è fatto apposta."

da *I Promessi Sposi* (1826) di A. Manzoni (1785–1873)

"This conclusion, though written by simple people, seems so apt that we have thought of putting it here, as the sauce for the whole story.
Which, if you did not dislike it, then show some sympathy to those who wrote it, and also a little to those who mended it. If, instead, you found it tiresome, believe that we did not do it on purpose."

from *The Betrothal* (1826) by A. Manzoni (1785–1873)

A. Metrology on planes, computing uncertainty in the homography

In section 4.1.1, in order to compute the planar homography $\mathtt{H}$, we seek the eigenvector $\mathbf{h}$ with smallest eigenvalue λ of the matrix $\mathtt{A}^\top\mathtt{A}$. If the measured points are noise-free, or $n = 4$, then $\mathbf{h} = \text{Null}(\mathtt{A})$, and in general it can be assumed that for $\mathbf{h}$ the residual error $\mathbf{h}^\top\mathtt{A}^\top\mathtt{A}\mathbf{h} = \lambda \approx 0$.

We now use matrix perturbation theory [50, 120, 134] to compute the covariance $\Lambda_{\mathbf{h}}$ of $\mathbf{h}$ based on this zero approximation. Let us define:

$$\mathbf{z} = \begin{pmatrix} x_1 \; y_1 \; x_2 \; y_2 \; x_3 \; y_3 \; \cdots \; x_n \; y_n \end{pmatrix}^\top$$

the vector of the $2n$ components of the n noisy image computation points, referred to an image coordinate system. Because of the noise we have:

$$\mathbf{z} = \tilde{\mathbf{z}} + \delta\mathbf{z} = \begin{pmatrix} \tilde{x}_1 \; \tilde{y}_1 \; \tilde{x}_2 \; \tilde{y}_2 \; \tilde{x}_3 \; \tilde{y}_3 \; \cdots \; \tilde{x}_n \; \tilde{y}_n \end{pmatrix}^\top + \begin{pmatrix} \delta x_1 \; \delta y_1 \; \delta x_2 \; \delta y_2 \; \delta x_3 \; \delta y_3 \; \cdots \; \delta x_n \; \delta y_n \end{pmatrix}^\top$$

where the $\tilde{}$ indicates noiseless quantities. Similarly for the world-plane computation points we define:

$$\mathbf{Z} = \begin{pmatrix} X_1 \; Y_1 \; X_2 \; Y_2 \; X_3 \; Y_3 \; \cdots \; X_n \; Y_n \end{pmatrix}^\top$$

the vector of the $2n$ components of the n noisy world-plane computation points referred to a world coordinate system and:

$$\mathbf{Z} = \tilde{\mathbf{Z}} + \delta\mathbf{Z} = \begin{pmatrix} \tilde{X}_1 \; \tilde{Y}_1 \; \tilde{X}_2 \; \tilde{Y}_2 \; \tilde{X}_3 \; \tilde{Y}_3 \; \cdots \; \tilde{X}_n \; \tilde{Y}_n \end{pmatrix}^\top + \begin{pmatrix} \delta X_1 \; \delta Y_1 \; \delta X_2 \; \delta Y_2 \; \delta X_3 \; \delta Y_3 \; \cdots \; \delta X_n \; \delta Y_n \end{pmatrix}^\top$$

We assume that the noise is Gaussian with null mean and that there is no correlation between the noise of different computation points. That means: $\text{E}(\delta z_i \delta z_j) = \delta_{ij}\sigma_i^2$ and $\text{E}(\delta Z_i \delta Z_j) = \delta_{ij}\Sigma_i^2$ where δ_{ij} is the Kronecker delta

$$\delta_{ij} = \begin{cases} 0 \; i \neq j \\ 1 \; i = j \end{cases}, i, j = 1 \cdots 2n$$

The generic odd row 1×9 vector of the $\mathtt{A}$ matrix is:

$$\mathbf{a}_{2i-1} = \begin{pmatrix} x_i \; y_i \; 1 \; 0 \; 0 \; 0 \; -x_i X_i \; -y_i X_i \; -X_i \end{pmatrix}$$

and the even row 1×9 vector is:

$$\mathbf{a}_{2i} = \begin{pmatrix} 0 \; 0 \; 0 \; x_i \; y_i \; 1 \; -x_i Y_i \; -y_i Y_i \; -Y_i \end{pmatrix}$$

with $i = 1 \cdots n$.

But $\mathbf{a}_{2i-1} = \tilde{\mathbf{a}}_{2i-1} + \delta\mathbf{a}_{2i-1}$ and $\mathbf{a}_{2i} = \tilde{\mathbf{a}}_{2i} + \delta\mathbf{a}_{2i}$ and therefore:

$$\delta\mathbf{a}_{2i-1} = \begin{pmatrix} \delta x_i \; \delta y_i \; 0\;0\;0\;0 \; -(\delta x_i X_i + x_i \delta X_i) \; -(\delta y_i X_i + y_i \delta X_i) \; -\delta X_i \end{pmatrix}$$

and:

$$\delta\mathbf{a}_{2i} = \begin{pmatrix} 0\;0\;0 \; \delta x_i \; \delta y_i \; 0 \; -(\delta x_i Y_i + x_i \delta Y_i) \; -(\delta y_i Y_i + y_i \delta Y_i) \; -\delta Y_i \end{pmatrix}$$

It is easy to prove that:

$$\left.\begin{array}{ll} \mathrm{E}(\delta\mathbf{a}_{2i-1}^\top \delta\mathbf{a}_{2j-1}) & = \delta_{ij}\mathtt{S}_i^o \\ \mathrm{E}(\delta\mathbf{a}_{2i}^\top \delta\mathbf{a}_{2j}) & = \delta_{ij}\mathtt{S}_i^e \\ \mathrm{E}(\delta\mathbf{a}_{2i-1}^\top \delta\mathbf{a}_{2j}) & = \delta_{ij}\mathtt{S}_i^{oe} \\ \mathrm{E}(\delta\mathbf{a}_{2i}^\top \delta\mathbf{a}_{2j-1}) & = \delta_{ij}\mathtt{S}_i^{eo} \end{array}\right\} \quad \forall i,j = 1\cdots n$$

where:

$$\mathtt{S}_i^o = \begin{bmatrix} \boldsymbol{\Lambda}_{\mathbf{x}_i} & 0 & -X_i\boldsymbol{\Lambda}_{\mathbf{x}_i} \\ 0 & 0 & 0 \\ -X_i\boldsymbol{\Lambda}_{\mathbf{x}_i} & 0 & X_i^2\boldsymbol{\Lambda}_{\mathbf{x}_i} + \Sigma_{X_i}^2 \mathbf{x}_i\mathbf{x}_i^\top \end{bmatrix}$$

$$\mathtt{S}_i^e = \begin{bmatrix} 0 & 0 & 0 \\ 0 & \boldsymbol{\Lambda}_{\mathbf{x}_i} & -Y_i\boldsymbol{\Lambda}_{\mathbf{x}_i} \\ 0 & -Y_i\boldsymbol{\Lambda}_{\mathbf{x}_i} & Y_i^2\boldsymbol{\Lambda}_{\mathbf{x}_i} + \Sigma_{Y_i}^2 \mathbf{x}_i\mathbf{x}_i^\top \end{bmatrix}$$

$$\mathtt{S}_i^{oe} = \begin{bmatrix} 0 & \boldsymbol{\Lambda}_{\mathbf{x}_i} & -Y_i\boldsymbol{\Lambda}_{\mathbf{x}_i} \\ 0 & 0 & 0 \\ 0 & -X_i\boldsymbol{\Lambda}_{\mathbf{x}_i} & X_iY_i\boldsymbol{\Lambda}_{\mathbf{x}_i} + \Sigma_{XY_i} \mathbf{x}_i\mathbf{x}_i^\top \end{bmatrix}, \qquad \mathtt{S}_i^{eo} = \mathtt{S}_i^{oe\top}$$

with $\boldsymbol{\Lambda}_{\mathbf{x}_i}$ the homogeneous covariance matrix of the image point $\mathbf{x}_i = (x_i, y_i, 1)^\top$ and $\boldsymbol{\Lambda}_{\mathbf{X}_i}$ the homogeneous covariance matrix of the world-point $\mathbf{X}_i$ in the form:

$$\boldsymbol{\Lambda}_{\mathbf{x}_i} = \begin{bmatrix} \sigma_{x_i}^2 & \sigma_{xy_i} & 0 \\ \sigma_{xy_i} & \sigma_{y_i}^2 & 0 \\ 0 & 0 & 0 \end{bmatrix}, \qquad \boldsymbol{\Lambda}_{\mathbf{X}_i} = \begin{bmatrix} \Sigma_{X_i}^2 & \Sigma_{XY_i} & 0 \\ \Sigma_{XY_i} & \Sigma_{Y_i}^2 & 0 \\ 0 & 0 & 0 \end{bmatrix}$$

In the case of homogeneous and isotropic noise, the above equations simplify.

Furthermore if we define the matrix $\mathtt{M} = \mathtt{A}^\top\mathtt{A}$ then:

$$\mathtt{M} = (\tilde{\mathtt{A}} + \delta\mathtt{A})^\top(\tilde{\mathtt{A}} + \delta\mathtt{A}) = \tilde{\mathtt{A}}^\top\tilde{\mathtt{A}} + \delta\mathtt{A}^\top\tilde{\mathtt{A}} + \tilde{\mathtt{A}}^\top\delta\mathtt{A} + \delta\mathtt{A}^\top\delta\mathtt{A}$$

Thus $\mathtt{M} = \tilde{\mathtt{M}} + \delta\mathtt{M}$ and for the first order approximation, $\delta\mathtt{M} = \delta\mathtt{A}^\top\tilde{\mathtt{A}} + \tilde{\mathtt{A}}^\top\delta\mathtt{A}$.

Now let us define $\mathbf{u}_1$ as the eigenvector corresponding to the null eigenvalue of the matrix $\tilde{\mathtt{M}}$ (the solution vector $\mathbf{h}$). The other eigensolutions are: $\tilde{\mathtt{M}}\tilde{\mathbf{u}}_j = \tilde{\lambda}_j\tilde{\mathbf{u}}_j$ with $j = 2\cdots 9$. It has been proven [50, 107] that the variation of the solution is related to the noise of the matrix as in the following formula:

$$\delta\mathbf{u}_1 = -\sum_{k=2}^{9} \frac{\tilde{\mathbf{u}}_k\tilde{\mathbf{u}}_k^\top}{\tilde{\lambda}_k}\delta\mathtt{M}\tilde{\mathbf{u}}_1$$

but $\delta\mathtt{M}\tilde{\mathbf{u}}_1 = \delta\mathtt{A}^\top\tilde{\mathtt{A}}\tilde{\mathbf{u}}_1 + \tilde{\mathtt{A}}^\top\delta\mathtt{A}\tilde{\mathbf{u}}_1$ and we know that $\tilde{\mathtt{A}}\tilde{\mathbf{u}}_1 = 0$, thus:

$$\delta\mathtt{M}\tilde{\mathbf{u}}_1 = \tilde{\mathtt{A}}^\top \delta\mathtt{A}\tilde{\mathbf{u}}_1$$

from where $\delta\mathbf{u}_1 = \tilde{\mathtt{J}}\tilde{\mathtt{A}}^\top \delta\mathtt{A}\tilde{\mathbf{u}}_1$ with $\tilde{\mathtt{J}} = -\sum_{k=2}^{9} \frac{\tilde{\mathbf{u}}_k \tilde{\mathbf{u}}_k^\top}{\tilde{\lambda}_k}$.

Therefore:

$$\Lambda_{\mathbf{h}} = \Lambda_{\mathbf{u}_1} = \mathrm{E}\left(\delta\mathbf{u}_1 \delta\mathbf{u}_1^\top\right) = \tilde{\mathtt{J}}\mathrm{E}\left(\tilde{\mathtt{A}}^\top \delta\mathtt{A}\tilde{\mathbf{u}}_1 \tilde{\mathbf{u}}_1^\top \delta\mathtt{A}^\top \tilde{\mathtt{A}}\right) \tilde{\mathtt{J}}^\top \tag{A.1}$$

$$= \tilde{\mathtt{J}}\mathrm{E}\left(\sum_{i=1}^{2n} \tilde{\mathbf{a}}_i^\top (\delta\tilde{\mathbf{a}}_i \cdot \tilde{\mathbf{u}}_1)(\sum_{j=1}^{2n} \tilde{\mathbf{a}}_j (\delta\tilde{\mathbf{a}}_j \cdot \tilde{\mathbf{u}}_1)\right) \tilde{\mathtt{J}}^\top$$

$$= \tilde{\mathtt{J}}\mathrm{E}\left(\sum_{i=1}^{2n} \tilde{\mathbf{a}}_i^\top (\sum_{j=1}^{2n} \tilde{\mathbf{a}}_j \tilde{\mathbf{u}}_1^\top (\delta\tilde{\mathbf{a}}_i^\top \delta\tilde{\mathbf{a}}_j)\tilde{\mathbf{u}}_1)\right) \tilde{\mathtt{J}}^\top$$

$$= \tilde{\mathtt{J}}\left(\sum_{i=1}^{2n} \tilde{\mathbf{a}}_i^\top (\sum_{j=1}^{2n} \tilde{\mathbf{a}}_j \tilde{\mathbf{u}}_1^\top \mathrm{E}(\delta\tilde{\mathbf{a}}_i^\top \delta\tilde{\mathbf{a}}_j)\tilde{\mathbf{u}}_1)\right) \tilde{\mathtt{J}}^\top$$

having used the fact that $(\delta\tilde{\mathbf{a}}_i \cdot \tilde{\mathbf{u}}_1)(\delta\tilde{\mathbf{a}}_j \cdot \tilde{\mathbf{u}}_1) = \tilde{\mathbf{u}}_1^\top (\delta\tilde{\mathbf{a}}_i^\top \delta\tilde{\mathbf{a}}_j)\tilde{\mathbf{u}}_1$.

Now considering that $\tilde{\mathtt{J}}$ is a symmetric matrix ($\tilde{\mathtt{J}}^\top = \tilde{\mathtt{J}}$) equation (A.1) can be written as:

$$\Lambda_{\mathbf{h}} = \tilde{\mathtt{J}}\mathtt{S}\tilde{\mathtt{J}}$$

where $\mathtt{S}$ is a 9×9 matrix obtained as follows:

$$\mathtt{S} = \sum_{i=1}^{n} (\tilde{\mathbf{a}}_{2i-1}^\top \tilde{\mathbf{a}}_{2i-1} \tilde{\mathbf{u}}_1^\top \mathrm{E}(\delta\tilde{\mathbf{a}}_{2i-1}^\top \delta\tilde{\mathbf{a}}_{2i-1})\tilde{\mathbf{u}}_1 + \tilde{\mathbf{a}}_{2i}^\top \tilde{\mathbf{a}}_{2i} \tilde{\mathbf{u}}_1^\top \mathrm{E}(\delta\tilde{\mathbf{a}}_{2i}^\top \delta\tilde{\mathbf{a}}_{2i})\tilde{\mathbf{u}}_1 +$$
$$\tilde{\mathbf{a}}_{2i-1}^\top \tilde{\mathbf{a}}_{2i} \tilde{\mathbf{u}}_1^\top \mathrm{E}(\delta\tilde{\mathbf{a}}_{2i-1}^\top \delta\tilde{\mathbf{a}}_{2i})\tilde{\mathbf{u}}_1 + \tilde{\mathbf{a}}_{2i}^\top \tilde{\mathbf{a}}_{2i-1} \tilde{\mathbf{u}}_1^\top \mathrm{E}(\delta\tilde{\mathbf{a}}_{2i}^\top \delta\tilde{\mathbf{a}}_{2i-1})\tilde{\mathbf{u}}_1)$$

yielding:

$$\mathtt{S} = \sum_{i=1}^{n} (\ \tilde{\mathbf{a}}_{2i-1}^\top \tilde{\mathbf{a}}_{2i-1} \tilde{\mathbf{u}}_1^\top \mathtt{S}_i^o \tilde{\mathbf{u}}_1 + \tilde{\mathbf{a}}_{2i}^\top \tilde{\mathbf{a}}_{2i} \tilde{\mathbf{u}}_1^\top \mathtt{S}_i^e \tilde{\mathbf{u}}_1 +$$
$$\tilde{\mathbf{a}}_{2i-1}^\top \tilde{\mathbf{a}}_{2i} \tilde{\mathbf{u}}_1^\top \mathtt{S}_i^{oe} \tilde{\mathbf{u}}_1 + \tilde{\mathbf{a}}_{2i}^\top \tilde{\mathbf{a}}_{2i-1} \tilde{\mathbf{u}}_1^\top \mathtt{S}_i^{eo} \tilde{\mathbf{u}}_1)$$

Note that many of the above equations require the true noise-free quantities, which in general are not available. Weng et al. [132] pointed out that if one writes, for instance, $\tilde{\mathtt{A}} = \mathtt{A} - \delta\mathtt{A}$ and substitutes this in the relevant equations, the term in $\delta\mathtt{A}$ disappears in the first order expression, allowing $\tilde{\mathtt{A}}$ to be simply interchanged with $\mathtt{A}$, and so on.

Finally the 9×9 covariance matrix $\Lambda_{\mathbf{h}}$ is:

$$\Lambda_{\mathbf{h}} = \mathtt{JSJ} \tag{A.2}$$

where $\mathtt{J} = -\sum_{k=2}^{9} \frac{\mathbf{u}_k \mathbf{u}_k^\top}{\lambda_k}$, with $\mathbf{u}_k$ the kth eigenvector of the $\mathtt{A}^\top\mathtt{A}$ matrix and λ_k the corresponding eigenvalue. The 9×9 matrix $\mathtt{S}$ is:

$$\mathtt{S} = \sum_{i=1}^{n} (\ \mathbf{a}_{2i-1}^\top \mathbf{a}_{2i-1} \mathbf{h}^\top \mathtt{S}_i^o \mathbf{h} + \mathbf{a}_{2i}^\top \mathbf{a}_{2i} \mathbf{h}^\top \mathtt{S}_i^e \mathbf{h} + \tag{A.3}$$
$$\mathbf{a}_{2i-1}^\top \mathbf{a}_{2i} \mathbf{h}^\top \mathtt{S}_i^{oe} \mathbf{h} + \mathbf{a}_{2i}^\top \mathbf{a}_{2i-1} \mathbf{h}^\top \mathtt{S}_i^{eo} \mathbf{h})$$

with $\mathbf{a}_i$ i^{th} row vector of the $\mathtt{A}$ matrix; n is the number of computation points.

B. Maximum likelihood estimation of end points for isotropic uncertainties

Given two image points $\mathbf{x}$ and $\mathbf{x}'$ with distributions $\Lambda_{\mathbf{x}}$ and $\Lambda_{\mathbf{x}'}$ isotropic but not necessarily equal (Fig. B.1), we estimate the points $\hat{\mathbf{x}}$ and $\hat{\mathbf{x}}'$ such that the cost function (5.24) is minimized and the alignment constraint (5.25) satisfied. It is a constrained minimization problem; a closed-form solution exists in this case.

The 2×2 covariance matrices $\Lambda_{\mathbf{x}}$ and $\Lambda_{\mathbf{x}'}$ for the two inhomogeneous end points $\mathbf{x}$ and $\mathbf{x}'$ define two circles with radius $r = \sigma_x = \sigma_y$ and $r' = \sigma_{x'} = \sigma_{y'}$ respectively. The line $\mathbf{l}$ through the vanishing point $\mathbf{v}$ that best fits the points $\mathbf{x}$ and $\mathbf{x}'$ can be computed as (in homogeneous form):

$$\mathbf{l} = \begin{pmatrix} 1 + \sqrt{1+\xi^2} \\ \xi \\ -(1+\sqrt{1+\xi^2})v_x - \xi v_y \end{pmatrix}, \quad \xi = 2\frac{r' d_x d_y + r d'_x d'_y}{r'(d_x^2 - d_y^2) + r(d_x'^2 - d_y'^2)}$$

where:

$$\mathbf{d} = \mathbf{x} - \mathbf{v}, \quad \mathbf{d}' = \mathbf{x}' - \mathbf{v}$$

Note that this formulation is valid if $\mathbf{v}$ is finite.

The required estimates $\hat{\mathbf{x}}$ and $\hat{\mathbf{x}}'$ are the orthogonal projections of the points $\mathbf{x}$ and $\mathbf{x}'$ onto the line $\mathbf{l}$:

$$\hat{\mathbf{x}} = \begin{pmatrix} l_y(\mathbf{x} \cdot \mathtt{F}\mathbf{l}) - l_x l_w \\ -l_x(\mathbf{x} \cdot \mathtt{F}\mathbf{l}) - l_y l_w \\ l_x^2 + l_y^2 \end{pmatrix}, \quad \hat{\mathbf{x}}' = \begin{pmatrix} l_y(\mathbf{x}' \cdot \mathtt{F}\mathbf{l}) - l_x l_w \\ -l_x(\mathbf{x}' \cdot \mathtt{F}\mathbf{l}) - l_y l_w \\ l_x^2 + l_y^2 \end{pmatrix} \tag{B.1}$$

with $\mathtt{F} = \begin{bmatrix} 0 & 1 & 0 \\ -1 & 0 & 0 \end{bmatrix}$.

The points $\hat{\mathbf{x}}$ and $\hat{\mathbf{x}}'$ obtained above are used to provide an initial solution in the general non-isotropic covariance case, for which a closed-form solution does not exist. In the general case, the non-isotropic covariance matrices $\Lambda_{\mathbf{x}}$ and $\Lambda_{\mathbf{x}'}$ are approximated by isotropic ones with radius

$$r = |det(\Lambda_{\mathbf{x}})|^{1/4} \quad r' = |det(\Lambda_{\mathbf{x}'})|^{1/4}$$

Then (B.1) is applied and the solution end points are refined by using a Levenberg-Marquardt numerical algorithm to minimize the cost function (5.24) while satisfying the alignment constraint (5.25).

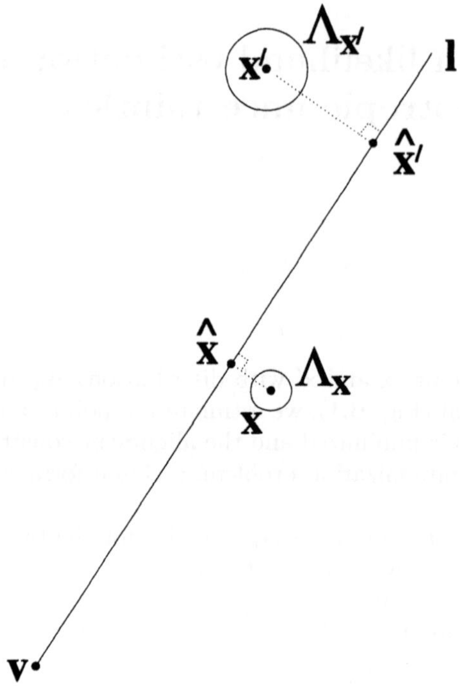

Fig. B.1. Maximum likelihood estimation of the location of two points subject to the alignment constraint. The uncertainty of the input points $\mathbf{x}$ and $\mathbf{x}'$ is Gaussian and isotropic. The points $\hat{\mathbf{x}}$ and $\hat{\mathbf{x}}'$ are the maximum likelihood estimates.

C. Single-view metrology, variance of distance between planes

C.1 Covariance of maximum likelihood estimation (MLE) end points

In Appendix B we have shown how to estimate the MLE points $\hat{\mathbf{x}}$ and $\hat{\mathbf{x}}'$ given two end points $\mathbf{x}$ and $\mathbf{x}'$ characterized by non-isotropic uncertainties. This section demonstrates how to compute the 4×4 covariance matrix of the MLE 4-vector $\hat{\boldsymbol{\zeta}} = \left(\hat{\mathbf{x}}^\top \; \hat{\mathbf{x}}'^\top \right)^\top$ from the covariances of the input points $\mathbf{x}$ and $\mathbf{x}'$ and the covariance of the projection matrix.

In order to simplify the following formulae we define the points: $\mathbf{b} = \mathbf{x}$ (referred to as the *base point*) on the plane π; and $\mathbf{t} = \mathbf{x}'$ (referred to as *top point*) on the plane π' corresponding to $\mathbf{x}$.

It can be shown that the 4×4 covariance matrix $\boldsymbol{\Lambda}_{\hat{\zeta}}$ of the vector $\hat{\boldsymbol{\zeta}} = \left(\hat{b}_x \; \hat{b}_y \; \hat{t}_x \; \hat{t}_y \right)^\top$ (MLE top and base points, see section 5.3.2) can be computed by using the *implicit function theorem* [20, 38] as:

$$\boldsymbol{\Lambda}_{\hat{\zeta}} = \mathtt{A}^{-1}\mathtt{B}\boldsymbol{\Lambda}_{\zeta}\mathtt{B}^\top\mathtt{A}^{-\top} \tag{C.1}$$

where $\boldsymbol{\zeta} = \left(b_x \; b_y \; t_x \; t_y \; p_{13} \; p_{23} \; p_{33} \right)^\top$ and:

$$\boldsymbol{\Lambda}_{\zeta} = \begin{bmatrix} \boldsymbol{\Lambda}_{\mathbf{b}} & 0 & 0 \\ 0 & \boldsymbol{\Lambda}_{\mathbf{t}} & 0 \\ 0 & 0 & \boldsymbol{\Lambda}_{\mathbf{p}_3} \end{bmatrix} \tag{C.2}$$

$\boldsymbol{\Lambda}_{\mathbf{b}}$ and $\boldsymbol{\Lambda}_{\mathbf{t}}$ are the 2×2 covariance matrices of the image points $\mathbf{b}$ and $\mathbf{t}$ respectively, and $\boldsymbol{\Lambda}_{\mathbf{p}_3}$ is the 3×3 covariance matrix of the vector $\mathbf{p}_3 = \alpha\mathbf{v}$ defined in (5.3). The assumption of statistical independence in (C.2) is a valid one.

The matrix $\mathtt{A}$ in (C.1) is the following 4×4 matrix:

$$\mathtt{A} = \begin{bmatrix} -\mathbf{e}_1^{\mathbf{b}} \cdot \boldsymbol{\delta}_t & -\mathbf{e}_2^{\mathbf{b}} \cdot \boldsymbol{\delta}_t & -\lambda p_{33}\delta_{t_y} & \lambda p_{33}\delta_{t_x} \\ \delta_{e_x}\delta_{b_y} & \delta_{e_y}\delta_{b_y} - \tau\lambda p_{33} & -\tau e^t_{11} - \lambda p_{33}\delta_{b_y} & -\tau e^t_{12} - \lambda p_{33}\delta_{b_y} \\ \tau\lambda p_{33} - \delta_{e_x}\delta_{b_x} & -\delta_{e_y}\delta_{b_x} & -\tau e^t_{12} + \lambda p_{33}\delta_{b_x} & -\tau e^t_{22} + \lambda p_{33}\delta_{b_x} \\ -\tau\delta_{t_y} & \tau\delta_{t_x} & \tau\delta_{b_y} & -\tau\delta_{b_x} \end{bmatrix}$$

where we have defined:

- $\mathtt{E}_{\mathbf{t}} = \boldsymbol{\Lambda}_{\mathbf{t}}^{-1}$ and e^t_{ij} its element in the (i,j) position;

- $\mathtt{E_b} = \boldsymbol{\Lambda}_{\mathbf{b}}^{-1}$ and $\mathbf{e}_1^{\mathbf{b}}$ and $\mathbf{e}_2^{\mathbf{b}}$ respectively its first and second row;
- $\mathbf{p} = (p_{13}, p_{23})^\top, \quad \boldsymbol{\delta}_{\mathbf{t}} = p_{33}\hat{\mathbf{t}} - \mathbf{p}, \quad \boldsymbol{\delta}_{\mathbf{b}} = p_{33}\hat{\mathbf{b}} - \mathbf{p}, \quad \boldsymbol{\delta}_{\mathbf{e}} = \mathbf{e}_2^{\mathbf{b}} - \mathbf{e}_1^{\mathbf{b}};$
- $\tau = (\mathbf{p}_3 \times \hat{\mathbf{t}})_y - (\mathbf{p}_3 \times \hat{\mathbf{t}})_x, \quad \lambda = \frac{\boldsymbol{\delta}_{\mathbf{e}}\cdot(\mathbf{b}-\hat{\mathbf{b}})}{\tau}.$

The matrix $\mathtt{B}$ in (C.1) is the following 4×7 matrix:

$$\mathtt{B} = \begin{bmatrix} \mathbf{e}_1^{\mathbf{b}} \cdot \boldsymbol{\delta}_t & \mathbf{e}_2^{\mathbf{b}} \cdot \boldsymbol{\delta}_t & 0 & 0 & \lambda\delta_{t_y} & -\lambda\delta_{t_x} & -\lambda\nu_1 \\ -\delta_{e_x}\delta_{b_y} & -\delta_{e_y}\delta_{b_y} & \tau e_{11}^t & \tau e_{12}^t & -\lambda\delta_{b_y} & -\lambda(\tau + \delta_{b_y}) & \lambda\nu_2 \\ \delta_{e_x}\delta_{b_x} & \delta_{e_y}\delta_{b_x} & \tau e_{12}^t & \tau e_{22}^t & \lambda(\tau + \delta_{b_x}) & \lambda\delta_{b_x} & -\lambda\nu_3 \\ 0 & 0 & 0 & 0 & \tau(\hat{t}_y - \hat{b}_y) & \tau(\hat{b}_x - \hat{t}_x) & \tau\nu_4 \end{bmatrix}$$

where we have defined:

$$\nu_1 = \hat{t}_y(p_{23}\hat{t}_x - p_{13}\hat{t}_y), \quad \nu_2 = \hat{b}_y(p_{13} + p_{23}) - p_{23}(\hat{t}_x + \hat{t}_y)$$

$$\nu_3 = \hat{b}_x(p_{13} + p_{23}) - p_{13}(\hat{t}_x + \hat{t}_y), \quad \nu_4 = \hat{t}_x\hat{b}_y - \hat{t}_y\hat{b}_x$$

If the vanishing point is noise-free then $\boldsymbol{\Lambda}_{\hat{\boldsymbol{\zeta}}}$ has rank 3, as expected because of the alignment constraint.

C.2 Variance of the distance measurement, σ_Z^2

As seen in section 5.3.2 the components of the vector $\hat{\boldsymbol{\zeta}}$ are used to compute the distance Z from (5.8) rewritten here as:

$$Z = -\frac{\|\hat{\mathbf{b}} \times \hat{\mathbf{t}}\|}{(\mathbf{p}_4 \cdot \hat{\mathbf{b}})\|\mathbf{p}_3 \times \hat{\mathbf{t}}\|}$$

with the MLE points $\hat{\mathbf{b}}$, $\hat{\mathbf{t}}$ homogeneous with unit third coordinate.

The variance σ_Z^2 of the measurement Z depends on the covariance of the $\hat{\boldsymbol{\zeta}}$ vector and the covariance of the 6-vector $\mathbf{p} = \left(\mathbf{p}_3^\top\ \mathbf{p}_4^\top\right)^\top$ computed in section 5.3.1. If $\hat{\boldsymbol{\zeta}}$ and $\mathbf{p}$ are statistically independent, then from first order error analysis:

$$\sigma_Z^2 = \nabla_Z \begin{pmatrix} \boldsymbol{\Lambda}_{\hat{\boldsymbol{\zeta}}} & 0 \\ 0 & \boldsymbol{\Lambda}_{\mathbf{p}} \end{pmatrix} \nabla_Z^\top \qquad \text{(C.3)}$$

the 1×10 Jacobian ∇_Z is:

$$\nabla_Z = Z \begin{pmatrix} \mathtt{F}\left(\frac{(\hat{\mathbf{t}}\times\hat{\mathbf{b}})\times\hat{\mathbf{t}}}{\beta^2} - \frac{\mathbf{p}_4}{\rho}\right) \\ \mathtt{F}\left(\frac{(\hat{\mathbf{b}}\times\hat{\mathbf{t}})\times\hat{\mathbf{b}}}{\beta^2} - \frac{(\mathbf{p}_3\times\hat{\mathbf{t}})\times\mathbf{p}_3}{\gamma^2}\right) \\ \frac{(\mathbf{p}_3\times\hat{\mathbf{t}})\times\hat{\mathbf{t}}}{\gamma^2} \\ -\frac{\hat{\mathbf{b}}}{\rho} \end{pmatrix}^\top$$

where we have defined $\beta = \|\hat{\mathbf{b}}\times\hat{\mathbf{t}}\|$, $\gamma = \|\mathbf{p}_3\times\hat{\mathbf{t}}\|$, $\rho = \mathbf{p}_4\cdot\hat{\mathbf{b}}$ and $\mathtt{F} = \begin{bmatrix} 1\,0\,0 \\ 0\,1\,0 \end{bmatrix}$.

Note that the assumption of statistical independence in (C.3) is an approximation.

D. Single-view metrology, variance of the affine parameter α

In section 5.2.1 the affine parameter α is obtained by computing the eigenvector $\mathbf{s}$ with the smallest eigenvalue of the matrix $\mathtt{A}^\top\mathtt{A}$ (5.8). If the measured reference points are noise-free, or $n = 1$, then $\mathbf{s} = \text{Null}(\mathtt{A})$ and in general we can assume that for $\mathbf{s}$ the residual error $\mathbf{s}^\top\mathtt{A}^\top\mathtt{A}\mathbf{s} = \lambda \approx 0$.

As in Appendix A we now use matrix perturbation theory [50, 120, 134] to compute the covariance $\boldsymbol{\Lambda}_\mathbf{s}$ of the solution vector $\mathbf{s}$ based on this zero approximation.

Note that the ith row of the matrix $\mathtt{A}$ depends on the normalized vanishing line $\mathbf{l}$, on the vanishing point $\mathbf{v}$, on the reference end points $\mathbf{b}_i, \mathbf{t}_i$ and on reference distances Z_i. Uncertainty in any of those elements induces an uncertainty in the matrix $\mathtt{A}$ and therefore uncertainty in the final solution $\mathbf{s}$. Let us define the input vector:

$$\boldsymbol{\eta} = (l_x\ l_y\ l_w\ v_x\ v_y\ v_w\ Z_1\ t_{1_x}\ t_{1_y}\ b_{1_x}\ b_{1_y}\ Z_n\ t_{n_x}\ t_{n_y}\ b_{n_x}\ b_{n_y})^\top$$

which contains the plane vanishing line, the vanishing point and the $5n$ components of the n references. Because of noise we have:

$$\begin{aligned}\boldsymbol{\eta} = \tilde{\boldsymbol{\eta}} + \delta\boldsymbol{\eta} = &(\tilde{l}_x\ \tilde{l}_y\ \tilde{l}_w\ \tilde{v}_x\ \tilde{v}_y\ \tilde{v}_w\ \tilde{Z}_1\ \tilde{t}_{1_x}\ \tilde{t}_{1_y}\ \tilde{b}_{1_x}\ \tilde{b}_{1_y}\ \tilde{Z}_n\ \tilde{t}_{n_x}\ \tilde{t}_{n_y}\ \tilde{b}_{n_x}\ \tilde{b}_{n_y})^\top + \\ &(\delta l_x\ \delta l_y\ \delta l_w\ \delta v_x\ \delta v_y\ \delta v_w\ \delta Z_1\ \delta t_{1_x}\ \delta t_{1_y} \delta Z_n\ \delta t_{n_x}\ \delta t_{n_y}\ \delta b_{n_x}\ \delta b_{n_y})^\top\end{aligned}$$

where the $\tilde{}$ indicates noiseless quantities.

We assume that the noise is Gaussian with zero mean and also that different reference distances are uncorrelated. However, the rows of the $\mathtt{A}$ matrix are correlated by the presence of $\mathbf{v}$ and $\mathbf{l}$ in each of them.

The 1×2 row-vector of the design matrix $\mathtt{A}$ is (see section 5.2.1):

$$\mathbf{a}_i = \left(Z_i \rho_i \gamma_i\ \beta_i \right)$$

with $i = 1 \cdots n$. Because of the noise $\mathbf{a}_i = \tilde{\mathbf{a}}_i + \delta\mathbf{a}_i$ and

$$\delta\mathbf{a}_i = \left(\rho_i\gamma_i\delta Z_i + Z_i\gamma_i\delta\rho_i + Z_i\rho_i\delta\gamma_i \quad \delta\beta_i \right)$$

It can be shown that $\delta\rho_i$, $\delta\gamma_i$ and $\delta\beta_i$ can be computed as functions of $\delta\boldsymbol{\eta}$ and, therefore, taking account of the statistical dependence of the rows of the $\mathtt{A}$ matrix, the 2×2 matrices $\mathtt{E}_{ij} = \text{E}(\delta\mathbf{a}_i^\top\delta\mathbf{a}_j)\ \forall i, j = 1 \cdots n$ can be computed.

Furthermore if we define the matrix $\mathtt{M} = \mathtt{A}^\top\mathtt{A}$ then:

$$\mathtt{M} = (\tilde{\mathtt{A}} + \delta\mathtt{A})^\top(\tilde{\mathtt{A}} + \delta\mathtt{A}) = \tilde{\mathtt{A}}^\top\tilde{\mathtt{A}} + \delta\mathtt{A}^\top\tilde{\mathtt{A}} + \tilde{\mathtt{A}}^\top\delta\mathtt{A} + \delta\mathtt{A}^\top\delta\mathtt{A}$$

Thus $\mathtt{M} = \tilde{\mathtt{M}} + \delta\mathtt{M}$ and for the first order approximation $\delta\mathtt{M} = \delta\mathtt{A}^\top\tilde{\mathtt{A}} + \tilde{\mathtt{A}}^\top\delta\mathtt{A}$.

As already stated, the vector $\mathbf{s}$ is the eigenvector corresponding to the null eigenvalue of the matrix $\tilde{\mathtt{M}}$; the other eigensolution is: $\tilde{\mathtt{M}}\tilde{\mathbf{u}} = \tilde{\lambda}_2\tilde{\mathbf{u}}$ with $\tilde{\mathbf{u}}$ the second eigenvector of the $\mathtt{A}^\top\mathtt{A}$ matrix and $\tilde{\lambda}_2$ the corresponding eigenvalue.

It is proved in [50, 108] that the variation of the solutions is related to the noise of the matrix $\mathtt{M}$ as:

$$\delta\mathbf{s} = -\frac{\tilde{\mathbf{u}}\tilde{\mathbf{u}}^\top}{\tilde{\lambda}_2}\delta\mathtt{M}\tilde{\mathbf{s}}$$

but since $\delta\mathtt{M}\tilde{\mathbf{s}} = \delta\mathtt{A}^\top\tilde{\mathtt{A}}\tilde{\mathbf{s}} + \tilde{\mathtt{A}}^\top\delta\mathtt{A}\tilde{\mathbf{s}}$ and $\tilde{\mathtt{A}}\tilde{\mathbf{s}} = 0$ then $\delta\mathtt{M}\tilde{\mathbf{s}} = \tilde{\mathtt{A}}^\top\delta\mathtt{A}\tilde{\mathbf{s}}$ and thus:

$$\delta\mathbf{s} = \tilde{\mathtt{J}}\tilde{\mathtt{A}}^\top\delta\mathtt{A}\tilde{\mathbf{s}}$$

where $\tilde{\mathtt{J}}$ is:

$$\tilde{\mathtt{J}} = -\frac{\tilde{\mathbf{u}}\tilde{\mathbf{u}}^\top}{\tilde{\lambda}_2}$$

Therefore:

$$\begin{aligned}
\Lambda_{\mathbf{s}} &= \mathrm{E}\left[\delta\mathbf{s}\delta\mathbf{s}^\top\right] = \tilde{\mathtt{J}}\mathrm{E}\left[\tilde{\mathtt{A}}^\top\delta\mathtt{A}\tilde{\mathbf{s}}\tilde{\mathbf{s}}^\top\delta\mathtt{A}^\top\tilde{\mathtt{A}}\right]\tilde{\mathtt{J}}^\top && \text{(D.1)}\\
&= \tilde{\mathtt{J}}\mathrm{E}\left[\sum_{i=1}^{n}\tilde{\mathbf{a}}_i^\top(\delta\tilde{\mathbf{a}}_i\cdot\tilde{\mathbf{s}})\sum_{j=1}^{n}\tilde{\mathbf{a}}_j(\delta\tilde{\mathbf{a}}_j\cdot\tilde{\mathbf{s}})\right]\tilde{\mathtt{J}}^\top\\
&= \tilde{\mathtt{J}}\mathrm{E}\left[\sum_{i=1}^{n}\tilde{\mathbf{a}}_i^\top\left(\sum_{j=1}^{n}\tilde{\mathbf{a}}_j\tilde{\mathbf{s}}^\top(\delta\tilde{\mathbf{a}}_i^\top\delta\tilde{\mathbf{a}}_j)\tilde{\mathbf{s}}\right)\right]\tilde{\mathtt{J}}^\top\\
&= \tilde{\mathtt{J}}\left[\sum_{i=1}^{n}\tilde{\mathbf{a}}_i^\top\left(\sum_{j=1}^{n}\tilde{\mathbf{a}}_j\tilde{\mathbf{s}}^\top\mathrm{E}(\delta\tilde{\mathbf{a}}_i^\top\delta\tilde{\mathbf{a}}_j)\tilde{\mathbf{s}}\right)\right]\tilde{\mathtt{J}}^\top
\end{aligned}$$

making use of the fact that:

$$(\delta\tilde{\mathbf{a}}_i\cdot\tilde{\mathbf{s}})(\delta\tilde{\mathbf{a}}_j\cdot\tilde{\mathbf{s}}) = \tilde{\mathbf{s}}^\top(\delta\tilde{\mathbf{a}}_i^\top\delta\tilde{\mathbf{a}}_j)\tilde{\mathbf{s}}$$

Now considering that $\tilde{\mathtt{J}}$ is a symmetric matrix ($\tilde{\mathtt{J}}^\top = \tilde{\mathtt{J}}$), equation (D.1) can be written as:

$$\Lambda_{\mathbf{s}} = \tilde{\mathtt{J}}\tilde{\mathtt{S}}\tilde{\mathtt{J}}$$

where $\tilde{\mathtt{S}}$ is the following 2×2 matrix:

$$\tilde{\mathtt{S}} = \sum_{i=1}^{n}\tilde{\mathbf{a}}_i^\top\left(\sum_{j=1}^{n}\tilde{\mathbf{a}}_j\tilde{\mathbf{s}}^\top\tilde{\mathtt{E}}_{ij}\tilde{\mathbf{s}}\right)$$

with $\tilde{\mathtt{E}}_{ij} = E(\delta\tilde{\mathbf{a}}_i^\top\delta\tilde{\mathbf{a}}_j)$.

Note that many of the above equations require the true noise-free quantities, which in general are not available. As in Appendix A, if one writes, for instance, $\tilde{\mathtt{A}} = \mathtt{A} - \delta\mathtt{A}$ and substitutes this in the relevant equations, the term in $\delta\mathtt{A}$

disappears in the first order expression, allowing $\tilde{\mathtt{A}}$ to be simply interchanged with $\mathtt{A}$, and so on. Therefore the 2×2 covariance matrix $\boldsymbol{\Lambda}_{\mathbf{s}}$ is simply:

$$\boldsymbol{\Lambda}_{\mathbf{s}} = \mathtt{JSJ} \tag{D.2}$$

where $\mathtt{J} = -\frac{\mathbf{u}\mathbf{u}^{\top}}{\lambda_2}$. The 2×2 matrix $\mathtt{S}$ is:

$$\mathtt{S} = \sum_{i=1}^{n} \mathbf{a}_i^{\top} \left(\sum_{j=1}^{n} \mathbf{a}_j \mathbf{s}^{\top} \mathtt{E}_{ij} \mathbf{s} \right) \tag{D.3}$$

with $\mathbf{a}_i$ the ith 1×2 row-vector of the design matrix $\mathtt{A}$ and n the number of references.

The 2×2 covariance matrix $\boldsymbol{\Lambda}_{\mathbf{s}}$ of the vector $\mathbf{s}$ is therefore computed.

Noise-free $\mathbf{v}$ *and* $\mathbf{l}$. In the case $\boldsymbol{\Lambda}_{\mathbf{l}} = \mathtt{0}$ and $\boldsymbol{\Lambda}_{\mathbf{v}} = \mathtt{0}$, so (D.3) simply becomes:

$$\mathtt{S} = \sum_{i=1}^{n} \mathbf{a}_i^{\top} \mathbf{a}_i \mathbf{s}^{\top} \mathtt{E}_{ii} \mathbf{s} \tag{D.4}$$

In fact the rows of the $\mathtt{A}$ matrix are all statistically independent.

Variance of α. It is easy to convert the 2×2 homogeneous covariance matrix $\boldsymbol{\Lambda}_{\mathbf{s}}$ in equation (D.2) into inhomogeneous coordinates. In fact, since $\mathbf{s} = \begin{pmatrix} s(1) & s(2) \end{pmatrix}^{\top}$ and $\alpha = \frac{s(1)}{s(2)}$ for a first order error analysis, the variance of the affine parameter α is:

$$\sigma_{\alpha}^2 = \boldsymbol{\nabla}\alpha \boldsymbol{\Lambda}_{\mathbf{s}} \boldsymbol{\nabla}\alpha^{\top} \tag{D.5}$$

with $\boldsymbol{\nabla}\alpha$ the 1×2 Jacobian matrix:

$$\boldsymbol{\nabla}\alpha = \frac{1}{s(2)^2} \begin{pmatrix} s(2) & -s(1) \end{pmatrix}$$

E. Metrology from planar parallax, derivations

E.1 Computing the motion constraints

In this section we derive some of the motion constraints described in Chapter 6. Extensive use is made of the lemma given in Fig. E.1.

Equation (6.4):

$$A_{\mathbf{p}_1\mathbf{p}_2\mathbf{x}_1} A_{\mathbf{q}_1\mathbf{q}_2\mathbf{p}_1} A_{\mathbf{x}_1\mathbf{x}_2\mathbf{q}_1} = A_{\mathbf{p}_1\mathbf{p}_2\mathbf{q}_1} A_{\mathbf{q}_1\mathbf{q}_2\mathbf{x}_1} A_{\mathbf{x}_1\mathbf{x}_2\mathbf{p}_1}$$

Proof. In Fig. 6.4 clearly we have:

$$\frac{A_{\mathbf{e}\mathbf{p}_1\mathbf{x}_1}}{A_{\mathbf{e}\mathbf{p}_1\mathbf{q}_1}} = \frac{A_{\mathbf{e}\mathbf{q}_1\mathbf{x}_1}}{A_{\mathbf{e}\mathbf{p}_1\mathbf{q}_1}} \frac{A_{\mathbf{e}\mathbf{p}_1\mathbf{x}_1}}{A_{\mathbf{e}\mathbf{q}_1\mathbf{x}_1}}$$

However by the lemma (substituting $\mathbf{a} \to \mathbf{p}_1$, $\mathbf{b} \to \mathbf{p}_2$, $\mathbf{c} \to \mathbf{q}_1$, $\mathbf{d} \to \mathbf{x}_1$)

$$\frac{A_{\mathbf{e}\mathbf{p}_1\mathbf{x}_1}}{A_{\mathbf{e}\mathbf{p}_1\mathbf{q}_1}} = \frac{A_{\mathbf{p}_1\mathbf{p}_2\mathbf{x}_1}}{A_{\mathbf{p}_1\mathbf{p}_2\mathbf{q}_1}}, \qquad \frac{A_{\mathbf{e}\mathbf{q}_1\mathbf{x}_1}}{A_{\mathbf{e}\mathbf{q}_1\mathbf{p}_1}} = \frac{A_{\mathbf{q}_1\mathbf{q}_2\mathbf{x}_1}}{A_{\mathbf{q}_1\mathbf{q}_2\mathbf{p}_1}}, \qquad \frac{A_{\mathbf{e}\mathbf{p}_1\mathbf{x}_1}}{A_{\mathbf{e}\mathbf{q}_1\mathbf{x}_1}} = \frac{A_{\mathbf{x}_1\mathbf{x}_2\mathbf{p}_1}}{A_{\mathbf{x}_1\mathbf{x}_2\mathbf{q}_1}}$$

Hence combining the identities above, we obtain the desired result:

$$A_{\mathbf{p}_1\mathbf{p}_2\mathbf{x}_1} A_{\mathbf{q}_1\mathbf{q}_2\mathbf{p}_1} A_{\mathbf{x}_1\mathbf{x}_2\mathbf{q}_1} = A_{\mathbf{p}_1\mathbf{p}_2\mathbf{q}_1} A_{\mathbf{q}_1\mathbf{q}_2\mathbf{x}_1} A_{\mathbf{x}_1\mathbf{x}_2\mathbf{p}_1}$$

□

Equation (6.1):

$$A_{\mathbf{p}_1\mathbf{p}_2\mathbf{x}_2} A_{\mathbf{q}_1\mathbf{q}_2\mathbf{x}_1} = A_{\mathbf{p}_1\mathbf{p}_2\mathbf{x}_1} A_{\mathbf{q}_1\mathbf{q}_2\mathbf{x}_2}$$

Proof. Applying the lemma to Fig. 6.4 we can write:

$$\frac{A_{\mathbf{p}_1\mathbf{p}_2\mathbf{x}_2}}{A_{\mathbf{p}_1\mathbf{p}_2\mathbf{x}_1}} = \frac{A_{\mathbf{e}\mathbf{x}_2\mathbf{p}_1}}{A_{\mathbf{e}\mathbf{x}_1\mathbf{p}_1}}, \qquad \frac{A_{\mathbf{q}_1\mathbf{q}_2\mathbf{x}_2}}{A_{\mathbf{q}_1\mathbf{q}_2\mathbf{x}_1}} = \frac{A_{\mathbf{e}\mathbf{x}_2\mathbf{q}_1}}{A_{\mathbf{e}\mathbf{x}_1\mathbf{q}_1}}$$

Also from the lemma:

$$\frac{A_{\mathbf{e}\mathbf{x}_2\mathbf{p}_1}}{A_{\mathbf{e}\mathbf{x}_2\mathbf{q}_1}} = \frac{A_{\mathbf{e}\mathbf{x}_1\mathbf{p}_1}}{A_{\mathbf{e}\mathbf{x}_1\mathbf{q}_1}}$$

Combining the above yields:

$$A_{\mathbf{p}_1\mathbf{p}_2\mathbf{x}_2} A_{\mathbf{q}_1\mathbf{q}_2\mathbf{x}_1} = A_{\mathbf{p}_1\mathbf{p}_2\mathbf{x}_1} A_{\mathbf{q}_1\mathbf{q}_2\mathbf{x}_2}$$

□

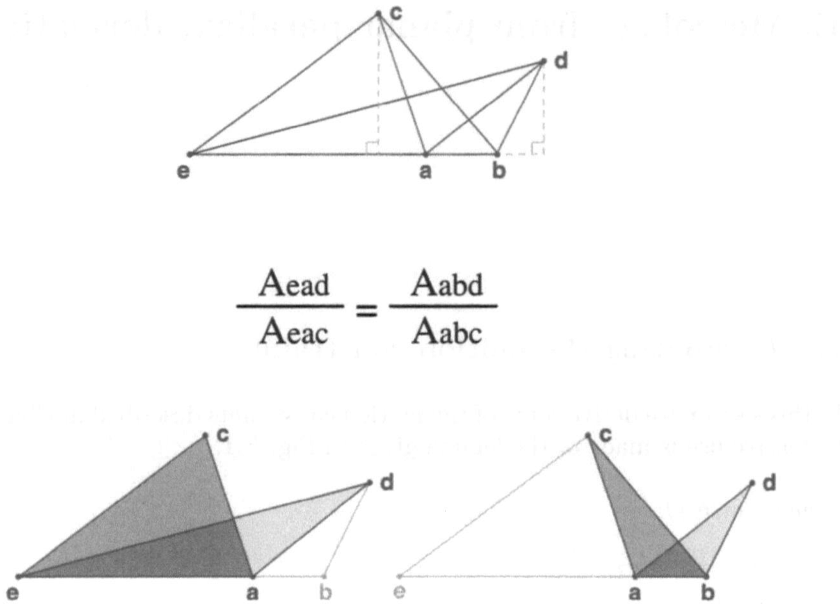

Fig. E.1. Lemma: The equality of the area ratios is easily seen by considering the relative heights and base lengths of the triangles.

E.2 Computing distances of points from planes directly in the images

In this section we show how we obtain (6.39) from (6.38).

First of all we need to compute the relationship between the area of a triangle on the distinguished plane and the area of its image.

Areas of triangles on the images.
For a generic triangle on the reference plane, the relation between its area and the area of its image on the first image can be computed as follows.

Let us assume that $\mathtt{H}_1$ and $\mathtt{H}_2$ are the image-to-world homographies for the first and second image respectively, such that:

$$\mathbf{X} = \mathtt{H}_1\mathbf{x} \quad \mathbf{X} = \mathtt{H}_2\mathbf{x}'$$

for every point $\mathbf{X}$ on the distinguished plane $\boldsymbol{\pi}$. Consequently $\mathbf{x} = \mathtt{H}\mathbf{x}'$, where:

$$\mathtt{H} = \mathtt{H}_1^{-1}\mathtt{H}_2$$

is the inter-image homography.

Then for the triangle of vertices $\mathbf{p}_1, \mathbf{p}_2, \mathbf{x}_1$, on the world plane we have:

$$\begin{aligned} A_{\mathbf{p}_1\mathbf{p}_2\mathbf{x}_1} &= \frac{|\mathtt{H}_1\mathbf{p} \;\; \mathtt{H}_2\mathbf{p}' \;\; \mathtt{H}_1\mathbf{x}|}{(\mathbf{l}\cdot\mathbf{p})(\mathbf{l}'\cdot\mathbf{p}')(\mathbf{l}\cdot\mathbf{x})} \\ &= \frac{|\mathtt{H}_1|}{(\mathbf{l}\cdot\mathbf{p})(\mathbf{l}'\cdot\mathbf{p}')(\mathbf{l}\cdot\mathbf{x})}|\mathbf{p} \;\; \mathtt{H}\mathbf{p}' \;\; \mathbf{x}| \end{aligned} \tag{E.1}$$

$$= \frac{|\mathtt{H}_1|}{(\mathbf{l}\cdot\mathbf{p})(\mathbf{l}'\cdot\mathbf{p}')(\mathbf{l}\cdot\mathbf{x})} A_{\mathbf{p}\dot{\mathbf{p}}\mathbf{x}}$$

with $\dot{\mathbf{p}} = \mathtt{H}\mathbf{p}'$. The area of the corresponding triangle on the first image is $A_{\mathbf{p}\dot{\mathbf{p}}\mathbf{x}}$.

Distances of points and cameras from the distinguished plane.
The distance of a point $\mathbf{X}$ from the plane can be computed if both image-to-world homographies are known by applying (6.38); but it can also be computed directly in the images if the inter-image homography $\mathtt{H}$ is known and the plane vanishing line $\mathbf{l}$ in either image is known. By applying (E.2) to (6.38), obtaining (6.39) is straightforward.

A similar procedure can be applied to compute the distance of the cameras in (6.35) from (6.34).

F. Metrology from planar parallax, variance of distances

As seen in section 6.4.2, the variance on the distance Z_x of a point $\mathbf{X}$ from a plane can be computed employing the plane-plus-parallax approach as:

$$\sigma^2_{Z_x} = \nabla_{Z_x} \Lambda \nabla^\top_{Z_x}$$

In this section we compute the 1×32 Jacobian ∇_{Z_x}.

F.1 Definitions

Equation (6.39) can be rewritten as:

$$Z_x = Z_p Z_q \frac{N}{D}$$

where we define:

- $N = N_1 - N_2$; $N_1 = \rho_q A_{\mathbf{x}\dot{\mathbf{x}}\dot{\mathbf{q}}} A_{\mathbf{p}\dot{\mathbf{p}}\dot{\mathbf{x}}}$; $N_2 = \rho_p A_{\mathbf{q}\dot{\mathbf{q}}\dot{\mathbf{x}}} A_{\mathbf{x}\dot{\mathbf{x}}\dot{\mathbf{p}}}$;
- $D = ZqD_1 + Z_p D_2$;
- $D_1 = A_{\mathbf{p}\dot{\mathbf{p}}\dot{\mathbf{x}}}(\rho_q A_{\mathbf{x}\dot{\mathbf{x}}\dot{\mathbf{q}}} - \rho_r A_{\mathbf{q}\dot{\mathbf{q}}\dot{\mathbf{x}}})$;
- $D_2 = A_{\mathbf{q}\dot{\mathbf{q}}\dot{\mathbf{x}}}(\rho_r A_{\mathbf{p}\dot{\mathbf{p}}\dot{\mathbf{x}}} - \rho_p A_{\mathbf{x}\dot{\mathbf{x}}\dot{\mathbf{p}}})$;
- $\rho_p = \mathbf{l} \cdot \mathbf{p}$; $\rho_q = \mathbf{l} \cdot \mathbf{q}$; $\rho_r = \mathbf{l} \cdot \mathbf{x}$.

For a generic triplet of points $\mathbf{a}, \mathbf{b}, \mathbf{c}$ we define: the determinant $D_{\mathbf{abc}} = \det \begin{bmatrix} \mathbf{a} & \mathbf{b} & \mathbf{c} \end{bmatrix}$; the area $A_{\mathbf{abc}} = |D_{\mathbf{abc}}|$; and the sign $S_{\mathbf{abc}} = \frac{D_{\mathbf{abc}}}{A_{\mathbf{abc}}}$.

Uncertainty is assumed on: the inter-image homography $\mathtt{H}$; the plane vanishing line $\mathbf{l}$; the two input references, their world distance and the position of the image points Z_p, $\mathbf{p}$, $\mathbf{p}'$, Z_q, $\mathbf{q}$, $\mathbf{q}'$; and the input image points $\mathbf{x}$, $\mathbf{x}'$.

F.2 Computing the Jacobian ∇_{Z_x}

Firstly we compute:

$$\frac{\partial Z_x}{\partial Z_p} = Z_q^2 \frac{N D_1}{D^2}$$

$$\frac{\partial Z_x}{\partial Z_q} = Z_p^2 \frac{N D_2}{D^2}$$

Furthermore for a generic vector $\mathbf{v}$ we have:

$$\frac{\partial Z_x}{\partial \mathbf{v}} = \frac{Z_p Z_q}{D^2}\left(\frac{\partial N}{\partial \mathbf{v}} D - N\left(Z_q \frac{\partial D_1}{\partial \mathbf{v}} + Z_p \frac{\partial D_2}{\partial \mathbf{v}}\right)\right)$$

where $\mathbf{v}$ can be $\mathbf{v} = \mathbf{l}, \mathbf{h}, \mathbf{p}, \mathbf{p}', \mathbf{q}, \mathbf{q}', \mathbf{x}, \mathbf{x}'$.

- Since $D_1 = A_{\mathbf{p}\dot{\mathbf{p}}\dot{\mathbf{x}}}(\rho_q A_{\mathbf{x}\dot{\mathbf{x}}\dot{\mathbf{q}}} - \rho_r A_{\mathbf{q}\dot{\mathbf{q}}\dot{\mathbf{x}}})$ we have:

$$\begin{aligned}
\frac{\partial D_1}{\partial \mathbf{l}} &= A_{\mathbf{p}\dot{\mathbf{p}}\dot{\mathbf{x}}}\left(A_{\mathbf{x}\dot{\mathbf{x}}\dot{\mathbf{q}}}\mathbf{q} - A_{\mathbf{q}\dot{\mathbf{q}}\dot{\mathbf{x}}}\mathbf{x}\right)\\
\frac{\partial D_1}{\partial \mathbf{h}} &= \frac{\partial A_{\mathbf{p}\dot{\mathbf{p}}\dot{\mathbf{x}}}}{\partial \mathbf{h}}\left(\rho_q A_{\mathbf{x}\dot{\mathbf{x}}\dot{\mathbf{q}}} - \rho_r A_{\mathbf{q}\dot{\mathbf{q}}\dot{\mathbf{x}}}\right) + A_{\mathbf{p}\dot{\mathbf{p}}\dot{\mathbf{x}}}\left(\rho_q \frac{\partial A_{\mathbf{x}\dot{\mathbf{x}}\dot{\mathbf{q}}}}{\partial \mathbf{h}} - \rho_r \frac{\partial A_{\mathbf{q}\dot{\mathbf{q}}\dot{\mathbf{x}}}}{\partial \mathbf{h}}\right)\\
\frac{\partial D_1}{\partial \mathbf{p}} &= \frac{D_1}{D_{\mathbf{p}\dot{\mathbf{p}}\dot{\mathbf{x}}}}(\dot{\mathbf{p}} \times \dot{\mathbf{x}})^\top\\
\frac{\partial D_1}{\partial \mathbf{p}'} &= \frac{D_1}{D_{\mathbf{p}\dot{\mathbf{p}}\dot{\mathbf{x}}}}(\dot{\mathbf{x}} \times \mathbf{p})^\top \mathtt{H}\\
\frac{\partial D_1}{\partial \mathbf{q}} &= A_{\mathbf{p}\dot{\mathbf{p}}\dot{\mathbf{x}}}\left(A_{\mathbf{x}\dot{\mathbf{x}}\dot{\mathbf{q}}}\mathbf{l}^\top - \rho_r S_{\mathbf{q}\dot{\mathbf{q}}\dot{\mathbf{x}}}(\dot{\mathbf{q}} \times \dot{\mathbf{x}})^\top\right)\\
\frac{\partial D_1}{\partial \mathbf{q}'} &= A_{\mathbf{p}\dot{\mathbf{p}}\dot{\mathbf{x}}}\left(\rho_q S_{\mathbf{x}\dot{\mathbf{x}}\dot{\mathbf{q}}}(\mathbf{x} \times \dot{\mathbf{x}})^\top - \rho_r S_{\mathbf{q}\dot{\mathbf{q}}\dot{\mathbf{x}}}(\dot{\mathbf{x}} \times \mathbf{q})^\top\right)\mathtt{H}\\
\frac{\partial D_1}{\partial \mathbf{x}} &= A_{\mathbf{p}\dot{\mathbf{p}}\dot{\mathbf{x}}}\left(\rho_q S_{\mathbf{x}\dot{\mathbf{x}}\dot{\mathbf{q}}}(\dot{\mathbf{x}} \times \dot{\mathbf{q}})^\top - A_{\mathbf{q}\dot{\mathbf{q}}\dot{\mathbf{x}}}\mathbf{l}^\top\right)\\
\frac{\partial D_1}{\partial \mathbf{x}'} &= \left(\frac{D_1}{D_{\mathbf{p}\dot{\mathbf{p}}\dot{\mathbf{x}}}}(\mathbf{p} \times \dot{\mathbf{p}})^\top + A_{\mathbf{p}\dot{\mathbf{p}}\dot{\mathbf{x}}}\left(\rho_q S_{\mathbf{x}\dot{\mathbf{x}}\dot{\mathbf{q}}}(\dot{\mathbf{q}} \times \mathbf{x})^\top - \rho_r S_{\mathbf{q}\dot{\mathbf{q}}\dot{\mathbf{x}}}(\mathbf{q} \times \dot{\mathbf{q}})^\top\right)\right)\mathtt{H}
\end{aligned}$$

- And since $D_2 = A_{\mathbf{q}\dot{\mathbf{q}}\dot{\mathbf{x}}}(\rho_r A_{\mathbf{p}\dot{\mathbf{p}}\dot{\mathbf{x}}} - \rho_p A_{\mathbf{x}\dot{\mathbf{x}}\dot{\mathbf{p}}})$ we have:

$$\begin{aligned}
\frac{\partial D_2}{\partial \mathbf{l}} &= A_{\mathbf{q}\dot{\mathbf{q}}\dot{\mathbf{x}}}\left(A_{\mathbf{p}\dot{\mathbf{p}}\dot{\mathbf{x}}}\mathbf{x} - A_{\mathbf{x}\dot{\mathbf{x}}\dot{\mathbf{p}}}\mathbf{p}\right)\\
\frac{\partial D_2}{\partial \mathbf{h}} &= \frac{\partial A_{\mathbf{q}\dot{\mathbf{q}}\dot{\mathbf{x}}}}{\partial \mathbf{h}}\left(\rho_r A_{\mathbf{p}\dot{\mathbf{p}}\dot{\mathbf{x}}} - \rho_p A_{\mathbf{x}\dot{\mathbf{x}}\dot{\mathbf{p}}}\right) + A_{\mathbf{q}\dot{\mathbf{q}}\dot{\mathbf{x}}}\left(\rho_r \frac{\partial A_{\mathbf{p}\dot{\mathbf{p}}\dot{\mathbf{x}}}}{\partial \mathbf{h}} - \rho_p \frac{\partial A_{\mathbf{x}\dot{\mathbf{x}}\dot{\mathbf{p}}}}{\partial \mathbf{h}}\right)\\
\frac{\partial D_2}{\partial \mathbf{p}} &= A_{\mathbf{q}\dot{\mathbf{q}}\dot{\mathbf{x}}}\left(\rho_r S_{\mathbf{p}\dot{\mathbf{p}}\dot{\mathbf{x}}}(\dot{\mathbf{p}} \times \dot{\mathbf{x}})^\top - A_{\mathbf{x}\dot{\mathbf{x}}\dot{\mathbf{p}}}\mathbf{l}^\top\right)\\
\frac{\partial D_2}{\partial \mathbf{p}'} &= A_{\mathbf{q}\dot{\mathbf{q}}\dot{\mathbf{x}}}\left(\rho_r S_{\mathbf{p}\dot{\mathbf{p}}\dot{\mathbf{x}}}(\dot{\mathbf{x}} \times \mathbf{p})^\top - \rho_p S_{\mathbf{x}\dot{\mathbf{x}}\dot{\mathbf{p}}}(\mathbf{x} \times \dot{\mathbf{x}})^\top\right)\mathtt{H}\\
\frac{\partial D_2}{\partial \mathbf{q}} &= \frac{D_2}{D_{\mathbf{q}\dot{\mathbf{q}}\dot{\mathbf{x}}}}(\dot{\mathbf{q}} \times \dot{\mathbf{x}})^\top\\
\frac{\partial D_2}{\partial \mathbf{q}'} &= \frac{D_2}{D_{\mathbf{q}\dot{\mathbf{q}}\dot{\mathbf{x}}}}(\dot{\mathbf{x}} \times \mathbf{q})^\top \mathtt{H}\\
\frac{\partial D_2}{\partial \mathbf{x}} &= A_{\mathbf{q}\dot{\mathbf{q}}\dot{\mathbf{x}}}\left(A_{\mathbf{p}\dot{\mathbf{p}}\dot{\mathbf{x}}}\mathbf{l}^\top - \rho_p S_{\mathbf{x}\dot{\mathbf{x}}\dot{\mathbf{p}}}(\dot{\mathbf{x}} \times \dot{\mathbf{p}})^\top\right)\\
\frac{\partial D_2}{\partial \mathbf{x}'} &= \left(\frac{D_2}{D_{\mathbf{q}\dot{\mathbf{q}}\dot{\mathbf{x}}}}(\mathbf{q} \times \dot{\mathbf{q}})^\top + A_{\mathbf{q}\dot{\mathbf{q}}\dot{\mathbf{x}}}\left(\rho_r S_{\mathbf{p}\dot{\mathbf{p}}\dot{\mathbf{x}}}(\mathbf{p} \times \dot{\mathbf{p}})^\top - \rho_p S_{\mathbf{x}\dot{\mathbf{x}}\dot{\mathbf{p}}}(\dot{\mathbf{p}} \times \mathbf{x})^\top\right)\right)\mathtt{H}
\end{aligned}$$

- Furthermore since $N = \rho_q A_{\mathbf{x}\dot{\mathbf{x}}\dot{\mathbf{q}}} A_{\mathbf{p}\dot{\mathbf{p}}\dot{\mathbf{x}}} - \rho_p A_{\mathbf{q}\dot{\mathbf{q}}\dot{\mathbf{x}}} A_{\mathbf{x}\dot{\mathbf{x}}\dot{\mathbf{p}}}$ we have:

$$
\begin{aligned}
\frac{\partial N}{\partial \mathbf{l}} &= A_{\mathbf{x}\dot{\mathbf{x}}\dot{\mathbf{q}}} A_{\mathbf{p}\dot{\mathbf{p}}\dot{\mathbf{x}}} \mathbf{q} - A_{\mathbf{q}\dot{\mathbf{q}}\dot{\mathbf{x}}} A_{\mathbf{x}\dot{\mathbf{x}}\dot{\mathbf{p}}} \mathbf{p} \\
\frac{\partial N}{\partial \mathbf{h}} &= \rho_q \left(A_{\mathbf{p}\dot{\mathbf{p}}\dot{\mathbf{x}}} \frac{\partial A_{\mathbf{x}\dot{\mathbf{x}}\dot{\mathbf{q}}}}{\partial \mathbf{h}} + A_{\mathbf{x}\dot{\mathbf{x}}\dot{\mathbf{q}}} \frac{\partial A_{\mathbf{p}\dot{\mathbf{p}}\dot{\mathbf{x}}}}{\partial \mathbf{h}} \right) - \rho_p \left(A_{\mathbf{x}\dot{\mathbf{x}}\dot{\mathbf{p}}} \frac{\partial A_{\mathbf{q}\dot{\mathbf{q}}\dot{\mathbf{x}}}}{\partial \mathbf{h}} + A_{\mathbf{q}\dot{\mathbf{q}}\dot{\mathbf{x}}} \frac{\partial A_{\mathbf{x}\dot{\mathbf{x}}\dot{\mathbf{p}}}}{\partial \mathbf{h}} \right) \\
\frac{\partial N}{\partial \mathbf{p}} &= \frac{N_1}{D_{\mathbf{p}\dot{\mathbf{p}}\dot{\mathbf{x}}}} (\dot{\mathbf{p}} \times \dot{\mathbf{x}})^\top - \frac{N_2}{\rho_p} \mathbf{l}^\top \\
\frac{\partial N}{\partial \mathbf{p}'} &= \left(\frac{N_1}{D_{\mathbf{p}\dot{\mathbf{p}}\dot{\mathbf{x}}}} (\dot{\mathbf{x}} \times \mathbf{p})^\top - \frac{N_2}{D_{\mathbf{x}\dot{\mathbf{x}}\dot{\mathbf{p}}}} (\mathbf{x} \times \dot{\mathbf{x}})^\top \right) \mathtt{H} \\
\frac{\partial N}{\partial \mathbf{q}} &= \frac{N_1}{\rho_q} \mathbf{l}^\top - \frac{N_2}{D_{\mathbf{q}\dot{\mathbf{q}}\dot{\mathbf{x}}}} (\dot{\mathbf{q}} \times \dot{\mathbf{x}})^\top \\
\frac{\partial N}{\partial \mathbf{q}'} &= \left(\frac{N_1}{D_{\mathbf{x}\dot{\mathbf{x}}\dot{\mathbf{q}}}} (\mathbf{x} \times \dot{\mathbf{x}})^\top - \frac{N_2}{D_{\mathbf{q}\dot{\mathbf{q}}\dot{\mathbf{x}}}} (\dot{\mathbf{x}} \times \mathbf{q})^\top \right) \mathtt{H} \\
\frac{\partial N}{\partial \mathbf{x}} &= \frac{N_1}{D_{\mathbf{x}\dot{\mathbf{x}}\dot{\mathbf{q}}}} (\dot{\mathbf{x}} \times \dot{\mathbf{q}})^\top - \frac{N_2}{D_{\mathbf{x}\dot{\mathbf{x}}\dot{\mathbf{p}}}} (\dot{\mathbf{x}} \times \dot{\mathbf{p}})^\top \\
\frac{\partial N}{\partial \mathbf{x}'} &= \left(\frac{N_1}{D_{\mathbf{x}\dot{\mathbf{x}}\dot{\mathbf{q}}}} (\dot{\mathbf{q}} \times \mathbf{x})^\top + \frac{N_1}{D_{\mathbf{p}\dot{\mathbf{p}}\dot{\mathbf{x}}}} (\mathbf{p} \times \dot{\mathbf{p}})^\top - \frac{N_2}{D_{\mathbf{q}\dot{\mathbf{q}}\dot{\mathbf{x}}}} (\mathbf{q} \times \dot{\mathbf{q}})^\top - \frac{N_2}{D_{\mathbf{x}\dot{\mathbf{x}}\dot{\mathbf{p}}}} (\dot{\mathbf{p}} \times \mathbf{x})^\top \right) \mathtt{H}
\end{aligned}
$$

Finally:

$$
\begin{aligned}
\frac{\partial A_{\mathbf{x}\dot{\mathbf{x}}\dot{\mathbf{p}}}}{\partial \mathbf{h}} &= S_{\mathbf{x}\dot{\mathbf{x}}\dot{\mathbf{p}}} \left((\dot{\mathbf{p}} \times \mathbf{x})^\top \mathtt{B}_{r'} + (\mathbf{x} \times \dot{\mathbf{x}})^\top \mathtt{B}_{p'} \right) \\
\frac{\partial A_{\mathbf{x}\dot{\mathbf{x}}\dot{\mathbf{q}}}}{\partial \mathbf{h}} &= S_{\mathbf{x}\dot{\mathbf{x}}\dot{\mathbf{q}}} \left((\dot{\mathbf{q}} \times \mathbf{x})^\top \mathtt{B}_{r'} + (\mathbf{x} \times \dot{\mathbf{x}})^\top \mathtt{B}_{q'} \right) \\
\frac{\partial A_{\mathbf{p}\dot{\mathbf{p}}\dot{\mathbf{x}}}}{\partial \mathbf{h}} &= S_{\mathbf{p}\dot{\mathbf{p}}\dot{\mathbf{x}}} \left((\dot{\mathbf{x}} \times \mathbf{p})^\top \mathtt{B}_{p'} + (\mathbf{p} \times \dot{\mathbf{p}})^\top \mathtt{B}_{r'} \right) \\
\frac{\partial A_{\mathbf{q}\dot{\mathbf{q}}\dot{\mathbf{x}}}}{\partial \mathbf{h}} &= S_{\mathbf{q}\dot{\mathbf{q}}\dot{\mathbf{x}}} \left((\dot{\mathbf{x}} \times \mathbf{q})^\top \mathtt{B}_{q'} + (\mathbf{q} \times \dot{\mathbf{q}})^\top \mathtt{B}_{r'} \right)
\end{aligned}
$$

where for a generic point $\mathbf{x}$:

$$
\mathtt{B}_x = \begin{bmatrix} \mathbf{x}^\top & \mathbf{0} & \mathbf{0} \\ \mathbf{0} & \mathbf{x}^\top & \mathbf{0} \\ \mathbf{0} & \mathbf{0} & \mathbf{x}^\top \end{bmatrix}
$$

The Jacobian ∇_{Z_x} is thus computed as:

$$
\nabla_{Z_x} = \left(\frac{\partial Z_x}{\partial \mathbf{l}} \; \frac{\partial Z_x}{\partial \mathbf{h}} \; \frac{\partial Z_x}{\partial Z_p} \; \frac{\partial Z_x}{\partial \mathbf{p}} \; \frac{\partial Z_x}{\partial \mathbf{p}'} \; \frac{\partial Z_x}{\partial Z_q} \; \frac{\partial Z_x}{\partial \mathbf{q}} \; \frac{\partial Z_x}{\partial \mathbf{q}'} \; \frac{\partial Z_x}{\partial \mathbf{x}} \; \frac{\partial Z_x}{\partial \mathbf{x}'} \right)
$$

- Furthermore since $N = \beta_e N_{\text{xxd}} l_{\text{out}} - \beta_p l_{\text{xxt}} l_{\text{xp}}$, we have:

$$\frac{\partial N}{\partial e} = l_{\text{xxd}} l_{\text{out}} - l_{\text{xxt}} l_{\text{xp}}$$

$$\frac{\partial N}{\partial u} = \beta_e \left(l_{\text{out}} \frac{\partial l_{\text{xxd}}}{\partial u} + l_{\text{xxd}} \frac{\partial l_{\text{out}}}{\partial u} \right) - \beta_p \left(l_{\text{xxt}} \frac{\partial l_{\text{out}}}{\partial u} + l_{\text{xp}} \frac{\partial l_{\text{xxt}}}{\partial u} \right)$$

$$\frac{\partial N}{\partial P} = \frac{N_1}{D_{\text{pnt}}} (\xi \times \mathbf{x}) - \frac{N_2}{\partial S} \mathbf{1}$$

$$\frac{\partial N}{\partial p} = \left(\frac{N_1}{D_{\text{pxx}}} (\mathbf{x} \times \mathbf{p}) + \frac{N_2}{D_{\text{xp}}} (\mathbf{x} \times \mathbf{x}) \right) \mathbf{R}$$

$$\frac{\partial N}{\partial q} = \frac{N_1}{l_{\text{q}}} \mathbf{1} - \frac{N_2}{D_{\text{qxx}}} (\mathbf{q} \times \xi)$$

$$\frac{\partial N}{\partial q} = \left(\frac{N_1}{D_{\text{xx}}} (\mathbf{x} \times \mathbf{x}) - \frac{N_2}{l_{\text{ers}}} (\mathbf{x} \times \mathbf{q})^{T} \right) \mathbf{R}$$

[illegible]

Finally

$$\frac{\partial \mathbf{E}}{\partial r} = \Sigma_{xx} \mathbf{H} (\mathbf{p} \times \mathbf{x})^{T} \mathbf{R}_{x} + (\mathbf{x} \times \mathbf{x})^{T} \mathbf{R}_{p}$$

[illegible]

where for any vector:

$$\mathbf{E}_{x} = \begin{bmatrix} 0 & & x \\ 0 & x & 0 \\ 0 & 0 & x \end{bmatrix}$$

The Jacobian ∇z is thus computed as:

$$\nabla z = \left([\text{illegible}] \right)$$

References

1. Leon Battista Alberti. *De Pictura.* 1435. Reproduced by Laterza (1980).
2. M Armstrong. Euclidean structure and camera calibration from image sequences. First year report, Dept. of Engineering Science, University of Oxford, 1994.
3. M Armstrong, A Zisserman and R Hartley. Self-calibration from image triplets. In *Proc. 4th European Conference on Computer Vision, Cambridge*, LNCS 1064/5, pages 3–16. Springer-Verlag, 1996.
4. S Avidan and A Shashua. Novel view synthesis in tensor space. In *Proceedings of the Conference on Computer Vision and Pattern Recognition*, 1997. Poster session 6.
5. P Beardsley, P Torr and A Zisserman. 3D model acquisition from extended image sequences. In *Proc. European Conference on Computer Vision*, LNCS 1064/1065, pages 683–695. Springer-Verlag, 1996.
6. P A Beardsley, D Sinclair and A Zisserman. Ego-motion from six points. Insight meeting, Catholic University Leuven, Feb. 1992.
7. M Berger. *Geometry II.* Springer-Verlag, 1987.
8. F Callari and F Ferrie. Active recognition: Using uncertainty to reduce ambiguity. Technical Report 11, Centre for Intelligent Machine, McGill University, Montreal, Quebec, Canada, September 1995.
9. J F Canny. Finding edges and lines in images. Master's thesis, MIT, 1983.
10. J F Canny. A computational approach to edge detection. *IEEE Transactions on Pattern Analysis and Machine Intelligence*, 8(6): 679–698, 1986.
11. D Capel and A Zisserman. Automated mosaicing with super-resolution zoom. In *Proceedings of the Conference on Computer Vision and Pattern Recognition, Santa Barbara*, pages 885–891, June 1998.
12. B Caprile and V Torre. Using vanishing points for camera calibration. *International Journal of Computer Vision*, 4:127–140, 1990.
13. S Carlsson. Duality of reconstruction and positioning from projective views. In *IEEE Workshop on Representation of Visual Scenes, Boston*, 1995.
14. S Carlsson and D Weinshall. Dual computation of projective shape and camera positions from multiple images. *International Journal of Computer Vision*, 27(3), 1998.
15. R Cipolla. *Active visual inference of surface shape.* PhD thesis, Dept. of Engineering Science, University of Oxford, 1991.
16. R Cipolla, T Drummond and D Robertson. Camera calibration from vanishing points in images of architectural scenes. In *Proc. 10th British Machine Vision Conference, Nottingham*, September 1999.
17. R Cipolla and P Giblin. *Visual Motion of Curves and Surfaces.* Cambridge University Press, October 1999. ISBN 0-521-63251-X.
18. R Cipolla, Y Okamoto and Y Kuno. Robust structure from motion using motion parallax. In *Proc. 4th International Conference on Computer Vision, Berlin*, pages 374–382, Los Alamitos, CA, 1993. IEEE Computer Society Press.
19. J C Clarke. First order error propagation: A primer. *Image and Vision Computing*, 1996.

20. J C Clarke. Modelling uncertainty: A primer. Technical Report 2161/98, University of Oxford, Dept. Engineering Science, 1998.
21. R T Collins and R S Weiss. Vanishing point calculation as a statistical inference on the unit sphere. In *Proc. 3rd International Conference on Computer Vision, Osaka*, pages 400–403, December 1990.
22. H Cramer. *Mathematical Methods of Statistics.* Princeton Univ. Press., 1946.
23. A Criminisi, I Reid and A Zisserman. Duality, rigidity and planar parallax. In *Proc. 5th European Conference on Computer Vision, Freiburg, Germany*, pages 846–861. Springer-Verlag, June 1998.
24. A Criminisi, I Reid and A Zisserman. A plane measuring device. *Image and Vision Computing*, 17(8): 625–634, 1999.
25. A Criminisi, I Reid and A Zisserman. Single view metrology. In *Proc. 7th International Conference on Computer Vision, Kerkyra, Greece*, pages 434–442, September 1999.
26. A Criminisi, I Reid and A Zisserman. Single view metrology. *International Journal of Computer Vision*, 40(2), November 2000. ISSN: 0920-5691.
27. A Criminisi and A Zisserman. Shape from texture: homogeneity revisited. In *Proc. 11th British Machine Vision Conference, Bristol*, pages 82–91, UK, September 2000.
28. A Criminisi, A Zisserman, L van Gool, S Bramble and D Compton. A new approach to obtain height measurements from video. In *Proc. of SPIE*, volume 3576, Boston, MA, 1-6 November 1998.
29. G Csurka, C Zeller, Z Zhang and O D Faugeras. Characterizing the uncertainty of the fundamental matrix. Technical Report 2560, I N R I A., France, 1995.
30. M De Mey. Perspektief in 3-D. *Gent Universiteit*, 8ste jaargang (3):14–17, December 1993.
31. P E Debevec, C J Taylor and J Malik. Modeling and rendering architecture from photographs: A hybrid geometry- and image-based approach. In *Proceedings, ACM SIGGRAPH*, pages 11–20, 1996.
32. Piero della Francesca. *De Prospectiva Pingendi.* Firenze, Italy, 1474. Reproduced by ed. Sansoni (1942), Edizione Critica.
33. F Devernay and O D Faugeras. Automatic calibration and removal of distortion from scenes of structured environments. In *Proc. of SPIE*, volume 2567, San Diego, CA, July 1995.
34. F Devernay and O D Faugeras. From projective to Euclidean reconstruction. In *Proceedings of the Conference on Computer Vision and Pattern Recognition*, pages 264–269, 1996.
35. S Dubin, J Nissanov, S Zietz, B Schrope, R Morano and R Hananiah. Bioengineering approach to non-invasive measurement of body composition. In *Rocky Mountain Bioengineering Symposium*, pages 21–23, April 1994.
36. Euclid. *Optica.* ca. 300 B. C. See also H E Burton, *The optics of Euclid* J Opt Soc Am 35, 357-372 (1945).
37. O Faugeras. What can be seen in three dimensions with an uncalibrated stereo rig? In *Proc. European Conference on Computer Vision*, LNCS 588, pages 563–578. Springer-Verlag, 1992.
38. O D Faugeras. *Three-Dimensional Computer Vision: a Geometric Viewpoint.* MIT Press, 1993.
39. O D Faugeras. Stratification of three-dimensional vision: projective, affine, and metric representation. *Journal of the Optical Society of America*, A12: 465–484, 1995.
40. O D Faugeras, S Laveau, L Robert, G Csurka and C Zeller. 3-D reconstruction of urban scenes from sequences of images. Tech. report, INRIA, 1995.
41. O D Faugeras, Q-T Luong, and S Maybank. Camera self-calibration: Theory and experiments. In *Proc. European Conference on Computer Vision*, LNCS 588, pages 321–334. Springer-Verlag, 1992.

42. O D Faugeras and T Papadopoulo. Grassmann–Cayley algebra for modeling systems of cameras and the algebraic equations of the manifold of trifocal tensors. Technical Report 3225, INRIA, Sophia-Antipolis, France, 1997.
43. O D Faugeras and L Robert. What can two images tell us about a third one. In J O Eckland, editor, *Proc. 3rd European Conference on Computer Vision, Stockholm*, pages 485–492. Springer-Verlag, 1994.
44. O D Faugeras and G Toscani. The calibration problem for stereo. In *Proceedings of the Conference on Computer Vision and Pattern Recognition*, pages 15–20, 1986.
45. J V Field. Alberti, the Abacus and Piero della Francesca's proof of perspective. *Renaissance Studies*, 11(2), 1997.
46. J V Field. *The Invention of Infinity, Mathematics and Arts in the Renaissance.* Oxford University Press, 1997.
47. J V Field, R Lunardi and T B Settle. *The Perspective Scheme of Masaccio's Trinity Fresco.* Leo S. Olschki Editore, Firenze, 1989.
48. F Figueroa and A Mahajan. A robust method to determine the coordinates of a wave source for 3-D position sensing. *ASME Journal of Dynamic Systems, Measurements and Control*, 116: 505–511, September 1994.
49. J J Gibson. *The Perception of the Visual World.* Houghton Mifflin, Boston, 1950.
50. G H Golub and C F Van Loan. *Matrix Computations.* The Johns Hopkins University Press, Baltimore, MD, second edition, 1989.
51. G Gracie. Analytical photogrammetry applied to single terrestrial photograph mensuration. In *XIth International Conference of Photogrammetry, Lausanne, Switzerland*, July 1968.
52. C J Harris and M Stephens. A combined corner and edge detector. In *Proc. 4th Alvey Vision Conference, Manchester*, pages 147–151, 1988.
53. R I Hartley. Estimation of relative camera positions for uncalibrated cameras. In *Proc. European Conference on Computer Vision*, LNCS 588, pages 579–587. Springer-Verlag, 1992.
54. R I Hartley. Lines and points in three views – a unified approach. In *ARPA Image Understanding Workshop, Monterrey*, 1994.
55. R I Hartley. Self-calibration from multiple views with a rotating camera. In *Proc. European Conference on Computer Vision*, LNCS 800/801, pages 471–478. Springer-Verlag, 1994.
56. R I Hartley. A linear method for reconstruction from lines and points. In *Proc. International Conference on Computer Vision*, pages 882–887, 1995.
57. R I Hartley and P Sturm. Triangulation. In *Proc. Conference Computer Analysis of Images and Patterns*, Prague, Czech Republic, 1995.
58. R I Hartley and A Zisserman. *Multiple View Geometry in Computer Vision.* Cambridge University Press, 2000, ISBN: 0521623049.
59. Y Horry, K Anjyo, and K Arai. Tour into the picture: Using a spidery mesh interface to make animation from a single image. In *Proceedings of the ACM SIGGRAPH Conference on Computer Graphics*, pages 225–232, 1997.
60. D Huynh. Affine reconstruction from monocular vision in the presence of a symmetry plane. In *Proc. 7th International Conference on Computer Vision, Kerkyra, Greece*, September 1999.
61. M Irani and P Anandan. Parallax geometry of pairs of points for 3D scene analysis. In B Buxton and R Cipolla, editors, *Proc. 4th European Conference on Computer Vision, LNCS 1064, Cambridge*, pages 17–30. Springer-Verlag, 1996.
62. M Irani, P Anandan and D Weinshall. From reference frames to reference planes: Multi-view parallax geometry and applications. In *Proc. European Conference on Computer Vision*, 1998.
63. A R Jarvis. A perspective on range finding techniques for computer vision. *IEEE Trans. on Pattern Analysis and Machine Intelligence*, 5(2): 122–139, 1983.

64. K Kanatani. Statistical optimization for geometric computation: theory and practice. Technical report, AI Lab, Dept of Computer Science, Gunma University, 1995.
65. S B Kang. Semiautomatic methods for recovering radial distortion parameters from a single image. Technical Report CRL 97/3, Digital CRL, 1997.
66. M Kass and A Witkin. Analyzing oriented patterns. In *Proc. International Joint Conference on Artificial Intelligence*, Los Angeles CA, 1985.
67. M Kemp. *The Science of Art.* Yale University Press, New Haven and London, 1989.
68. T Kim, Y Seo and K Hong. Physics-based 3D position analysis of a soccer ball from monocular image sequences. *Proc. International Conference on Computer Vision*, pages 721–726, 1998.
69. J J Koenderink and A J van Doorn. Affine structure from motion. *J. Opt. Soc. Am. A*, 8(2):377–385, 1991.
70. E Kruppa. Zur ermittlung eines objektes aus zwei perspektiven mit innerer orientierung. *Sitz.-Ber. Akad. Wiss., Wien, Math. Naturw. Abt. IIa*, 122:1939–1948, 1913.
71. R Kumar, P Anandan and K Hanna. Shape recovery from multiple views: a parallax based approach. In *ARPA Image Understanding Workshop, Monterey, CA*, 2929 Campus Drive, Suite 260, San Mateo, California 94403o, November 1994. ARPA, Image Understanding, Morgan Kauffmann Publishers.
72. R Kumar, P Anandan, M Irani, J Bergen and K Hanna. Representation of scenes from collections of images. In *ICCV Workshop on the Representation of Visual Scenes*, 1995.
73. R Kumar and A R Hanson. Robust methods for estimating pose and a sensitivity analysis. *Computer Vision, Graphics and Image Processing*, vol. 60(3): 313–342, 1994.
74. J M Lawn and R Cipolla. Epipole estimation using affine motion-parallax. In *Proc. 4th British Machine Vision Conference, Guildford*, 1993.
75. J M Lawn and R Cipolla. Robust egomotion estimation from affine motion parallax. In *Proc. 3rd European Conference on Computer Vision, Stockholm*, pages 205–210. Springer-Verlag, 1994.
76. M Levoy. The digital Michelangelo project. In *Proc. EuroGraphics*, volume 18, September 1999.
77. D Liebowitz, A Criminisi, and A Zisserman. Creating architectural models from images. In *Proc. EuroGraphics*, volume 18, pages 39–50, September 1999.
78. D Liebowitz and A Zisserman. Metric rectification for perspective images of planes. In *Proceedings of the Conference on Computer Vision and Pattern Recognition*, pages 482–488, June 1998.
79. H C Longuet-Higgins. A computer algorithm for reconstructing a scene from two projections. *Nature*, 293: 133–135, September 1981.
80. Q-T Luong and T Viéville. Canonic representations for the geometries of multiple projective views. In *Proc. 3rd European Conference on Computer Vision, Stockholm*, pages 589–599, May 1994.
81. H-G Maas. Robust automatic surface reconstruction with structured light. In *International Archives of Photogrammetry and Remote Sensing*, volume XXIX of *Part B5*, pages 102–107. 1992.
82. J Malik. Kerkyra, Greece, September 1999. Personal communication.
83. B Matei and P Meer. Bootstraping a heteroscedastic regression model with application to 3D rigid motion evaluation. In *Workshop on Vision Algorithms, Theory and Practice*, Kerkyra, Greece, September 1999. Springer-Verlag.
84. G F McLean and D Kotturi. Vanishing point detection by line clustering. *IEEE Transactions on Pattern Analysis and Machine Intelligence*, 17(11):1090–1095, 1995.

85. R Mohr. Projective geometry and computer vision. In C H Chen, L F Pau, and P S P Wang, editors, *Handbook of Pattern Recognition and Computer Vision.* World Scientific, 1992.
86. R Mohr, B Boufama and P Brand. Accurate projective reconstruction. In *Applications of Invariance in Computer Vision*, pages 257–276, Aores, Portugal, October 1993.
87. T Moons, L van Gool, M van Diest and A Oosterlinck. Affine structure from perspective image pairs obtained by a translating camera. In J L Mundy, A Zisserman and D Forsyth, editors, *Applications of invariance in computer vision*, pages 297–316. Springer-Verlag, 1994.
88. J Mundy and A Zisserman. *Geometric Invariance in Computer Vision.* MIT Press, 1992.
89. J Mundy, A Zisserman and D Forsyth. *Applications of Invariance in Computer Vision.* LNCS 825. Springer-Verlag, 1994.
90. S Nayar and S Narasimhan. Vision in bad weather. In *Proc. 7th International Conference on Computer Vision, Kerkyra, Greece*, pages 820–827, September 1999.
91. National Gallery of Scotland. *Dutch Church Painters.* The National Galleries of Scotland, Edinburgh, 1984.
92. S Peccatori and S Zuffi. *ArtBook Vermeer: The master of light and composition - his life in paintings.* Dorling Kindersley, 1999.
93. S B Pollard. *Identifying Correspondences in Binocular Stereo.* PhD thesis, University of Sheffield, 1985.
94. M Pollefeys, R Koch and L van Gool. Self calibration and metric reconstruction in spite of varying and unknown internal camera parameters. In *Proc. 6th International Conference on Computer Vision, Bombay, India*, pages 90–96, 1998.
95. W Press, B Flannery, S Teukolsky and W Vetterling. *Numerical Recipes in C.* Cambridge University Press, 1988.
96. M Proesmans, T Tuytelaars and L J van Gool. Monocular image measurements. Technical Report Improofs-M12T21/1/P, K.U.Leuven, 1998.
97. M Proesmans, L J van Gool, and F Defoort. Reading between the lines – a method for extracting dynamic 3D with texture. In *Proc. 6th International Conference on Computer Vision, Bombay, India*, pages 1081–1086, 1998.
98. L Quan and R Mohr. Affine shape representation from motion through reference points. *Journal of Mathematical Imaging and Vision*, 1:145–151, 1992.
99. I Reid and A Zisserman. Accurate metrology in uncalibrated video sequences. Technical report, Oxford University, Dept. of Engineering Science, 1996.
100. I Reid and A Zisserman. Goal-directed video metrology. In R Cipolla and B Buxton, editors, *Proc. 4th European Conference on Computer Vision, LNCS 1065, Cambridge*, volume II, pages 647–658. Springer-Verlag, April 1996.
101. I D Reid and A North. 3D trajectories from a single viewpoint using shadows. *Proc. British Machine Vision Conference*, 1998.
102. J A Rice. *Mathematical Statistics and Data Analysis.* Wadsworth and Brooks, California, 1988.
103. L Robert and O D Faugeras. Relative 3D positioning and 3D convex hull computation from a weakly calibrated stereo pair. In *Proc. 4th International Conference on Computer Vision, Berlin*, pages 540–544, 1993.
104. C Rothwell, D Forsyth, A Zisserman and J Mundy. Extracting projective structure from single perspective views of 3D point sets. In *Proc. International Conference on Computer Vision*, pages 573–582, 1993.
105. H S Sawhney. Simplifying motion and structure analysis using planar parallax and image warping. In *Proceedings of the Conference on Computer Vision and Pattern Recognition*, 1994.
106. J G Semple and G T Kneebone. *Algebraic Projective Geometry.* Oxford University Press, 1979.

107. L S Shapiro and J M Brady. Rejecting outliers and estimating errors in an orthogonal regression framework. Technical report 1974/93, Department of Engineering Science, University of Oxford, 1993.
108. L S Shapiro and J M Brady. Rejecting outliers and estimating errors in an orthogonal regression framework. *Philosophical Transactions of the Royal Society of London, Series A*, 350:407–439, 1995.
109. L S Shapiro, A Zisserman and M Brady. 3D motion recovery via affine epipolar geometry. *International Journal of Computer Vision*, 16(2): 147–182, 1995.
110. A Shashua. On geometric and algebraic aspects of 3D affine and projective structures from perspective 2D views. In J Mundy, A Zisserman and D Forsyth, editors, *Applications of Invariance in Computer Vision LNCS 825*, pages 127–143. Springer-Verlag, 1994.
111. A Shashua. Trilinearity in visual recognition by alignment. In *Proc. 3rd European Conference on Computer Vision, Stockholm*, volume 1, pages 479–484, May 1994.
112. A Shashua. Multiple-view geometry and photometry. In Springer-Verlag, editor, *ACCV*, Singapore, December 1995.
113. A Shashua. On photometric issues in 3D visual recognition from a single 2D image. International journal of computer vision, Hebrew University, 1995.
114. J A Shufelt. Performance and analysis of vanishing point detection techniques. *IEEE Transactions on Pattern Analysis and Machine Intelligence*, 21(3): 282–288, March 1999.
115. D A Sinclair. *Experiments in Motion and Correspondence.* PhD thesis, University of Oxford, 1992.
116. M E Spetsakis and J Aloimonos. Structure from motion using line correspondences. *International Journal of Computer Vision*, 4(3):171–183, 1990.
117. C E Springer. *Geometry and Analysis of Projective Spaces.* Freeman, 1964.
118. G P Stein. *Geometric and Photometric Constraints: Motion and Structure from three Views.* PhD thesis, MIT, 1997.
119. G Stern and A Schindler. Three-dimensional visualization of bone surfaces from ultrasound scanning. Technical report, A.I. DuPont Institute, 1994.
120. G W Stewart and J Sun. *Matrix Perturbation Theory.* Academic Press Inc., USA, 1990.
121. K A Stroud. *Engineering Mathematics.* MacMillan Education, 3rd edition, 1987.
122. B Taylor. *Linear Perspective.* London, 1715.
123. P H S Torr and A Zisserman. Robust parametrization and computation of the trifocal tensor. In *Proc. 7th British Machine Vision Conference, Edinburgh*, 1996.
124. W Triggs. Plane + parallax, tensors and factorization. In *Proc. 6th European Conference on Computer Vision, Dublin, Ireland*, 2000.
125. Y R Tsai. An efficient and accurate camera calibration technique for 3D machine vision. In *Proceedings of the Conference on Computer Vision and Pattern Recognition*, 1986.
126. Y R Tsai. A versatile camera calibration technique for high-accuracy 3D machine vision metrology using off-the-shelf TV cameras and lenses. *IEEE Journal of Robotics and Automation*, RA-3(4): 323–344, August 1987.
127. L van Gool, M Proesmans and A Zisserman. Grouping and invariants using planar homologies. In *Workshop on Geometrical Modeling and Invariants for Computer Vision.* Xidian University Press, 1995.
128. L van Gool, M Proesmans, and A Zisserman. Planar homologies as a basis for grouping and recognition. *Image and Vision Computing*, 16: 21–26, January 1998.
129. T Viéville and D Lingrand. Using singular displacements for uncalibrated monocular vision systems. Technical Report 2678, INRIA, 1995.
130. D Weinshall, P Anandan and M Irani. From ordinal to Euclidean reconstruction with partial scene calibration. In R Koch and L van Gool, editors, *Proc. of SMILE98, 3D Structure from Multiple Images of Large-Scale Environments*, pages 208–223. Springer-Verlag, 1998.

131. D Weinshall, M Werman, and A Shashua. Duality of multi-point and multi-frame geometry: Fundamental shape matrices and tensors. In B Buxton and R Cipolla, editors, *Proc. 4th European Conference on Computer Vision, LNCS 1065, Cambridge*, pages 217–227. Springer–Verlag, 1996.
132. J Weng, T S Huang and N Ahuja. Motion and structure from two perspective views: algorithms, error analysis and error estimation. *IEEE Transactions on Pattern Analysis and Machine Intelligence*, 11(5): 451–476, 1989.
133. M Werman, D Weinshall and A Shashua. Shape tensors for efficient and learnable indexing. In *IEEE Workshop on Representation of Visual Scenes, Boston*, 1995.
134. J H Wilkinson. *The Algebraic Eigenvalue Problem.* Clarendon Press, Oxford, 1965.
135. Z Zhang. Determining the epipolar geometry and its uncertainty: A review. *International Journal of Computer Vision*, 1997.
136. Z Zhang, P Anandan and H Y Shum. What can be determined from a full and a weak perspective image? In *Proc. 7th International Conference on Computer Vision, Kerkyra, Greece*, September 1999.
137. Z Zhang, R Deriche, O D Faugeras and Q-T Luong. A robust technique for matching two uncalibrated images through the recovery of the unknown epipolar geometry. *Artificial Intelligence*, 78: 87–119, 1995.
138. Z Zhang and A R Hanson. 3D reconstruction based on homography mapping. In *ARPA Image Understanding workshop*, Palm Springs, CA, 1996.
139. Z Zhang, Q-T Luong and O D Faugeras. Motion of an uncalibrated stereo rig: Self-calibration and metric reconstruction. *IEEE Trans. Robotics and Automation*, 12(1): 103–113, Feb. 1996.
140. A Zisserman, P Beardsley and I Reid. Metric calibration of a stereo rig. In *IEEE Workshop on Representation of Visual Scenes, Boston*, pages 93–100, 1995.

Index

MIX
Papier aus verantwortungsvollen Quellen
Paper from responsible sources
FSC® C105338

If you have any concerns about our products,
you can contact us on
ProductSafety@springernature.com

In case Publisher is established outside the EU,
the EU authorized representative is:
Springer Nature Customer Service Center GmbH
Europaplatz 3, 69115 Heidelberg, Germany

Printed by Libri Plureos GmbH
in Hamburg, Germany